MONTY HALLS'
GREAT ESCAPE

MONTY HALLS'
GREAT ESCAPE
BEACHCOMBER COTTAGE

My Search for the Simple Life

MONTY HALLS

BBC
BOOKS

Published to accompany the television series *Monty Halls' Great Escape*, produced by Tigress Productions and first broadcast on BBC Two in 2009. Executive producers: Dick Colhurst, Wendy Rattray and Emma Willis. Producer and director: Martin Pailthorpe.

10 9 8

First published in 2009 by BBC Books, an imprint of Ebury Publishing. A Random House Group Company.

The Random House Group Limited Reg. No. 954009.
Addresses for companies within the Random House Group can be found at www.randomhouse.co.uk

A CIP catalogue record for this book is available from the British Library.

ISBN 978 1 846 07621 3

The Random House Group Limited supports The Forest Stewardship Council (FSC), the leading international forest certification organisation. All our titles that are printed on Greenpeace approved FSC certified paper carry the FSC logo. Our paper procurement policy can be found at www.rbooks.co.uk/ environment

Commissioning editor: Albert DePetrillo
In-house editor: Christopher Tinker
Project editor: Steve Tribe
Illustrations: Patsy Breach
Production: Phil Spencer

Printed and bound in Great Britain by Clays of St Ives plc

To buy books by your favourite authors and register for offers, visit www.rbooks.co.uk

Contents

This book is dedicated to Robbie Stuart
- a proud Scotsman who fought like a lion
to the very end.

Introduction

There is a real temptation when writing about the west coast of Scotland to lapse into clichés. The entire place demands hyperbole – the landscape, the people, the wildlife, and the heritage. It all seems impossibly romantic upon arrival, and gets more dramatic the deeper one becomes immersed and absorbed. Particularly in a place like Applecross – surely one of the most beautiful locations in Britain (you see, I'm doing it now) – where every bend in the road reveals great landscapes, and where eagles shriek and stags leap with positively indecent frequency. In writing terms, I had run out of superlatives in the first week and had filled several notebooks with deranged babblings about the grandeur of the setting. The next stage could well have been wandering the hills in a baggy shirt and a kilt, periodically bursting into song about the heather and littering my conversation with the word 'Bonny'. I wasn't just seduced by the west coast; I positively eloped with it, showing slavish devotion in a one-sided relationship that could well have ended in tears. Happily, sanity prevailed, and I began to see the reality beyond the film-star looks, experiencing the highs and lows of any protracted stay in a single location. There are, however, things that must be said about Applecross, and I'm delighted to be given a brief opportunity to say them.

I travelled into the heart of the west coast trembling in anticipation. I knew it was going to be beautiful, but I harboured genuine fears about how I would be received by the local population. I was, after all, chasing a dream, head full of romantic notions, none of them backed up by the practical skills that are essential to surviving and prospering as a crofter. I was heading into the heart of a small, tight-knit community where I would be wholly reliant on the people around me. Although the idea of the quaint crofting and fishing village is an enchanting one, I had experienced enough small communities in my travels around the world to know that hostility and suspicion might well be the default setting upon my arrival. As I twisted down the road over the mountain pass with the village sitting on the coast below me, I had the feeling of dropping into the unknown, a realisation that the next six months could be a very lonely time indeed.

From the moment I pushed open the door of the pub on that first night, the people of Applecross showed me nothing but consideration, understanding, friendship and hospitality. The only prejudices that existed did so within myself, a city boy awaiting the first rip off, the first sneer, the first gesture of open hostility. It never materialised, and that was the making of my experience in the village. On one level there was the acquisition of new skills on a daily basis under the tutelage of a series of patient, skilled instructors, passing on their knowledge to someone who must have appeared positively childlike in the depth of his ignorance. Next was learning what community was all about, something we seem to have lost in the headlong rush of modern life. Finally, there was the setting itself, a tiny cottage on a windswept headland where the only noise came from the wind and the sea, a suitably dramatic backdrop for a time in my life that will always be special.

There was also the ever-present spectre of Gavin Maxwell, the inspiration for the journey. I was daunted by the prospect of slipping, albeit inadequately and briefly, into his shoes. I re-read his trilogy about his life with the otters at Camusfearna when I was up there, and commend these books. No one has captured the essence of the west coast quite like Maxwell, and I speak as a man who turned the pages in the gathering gloom of the bothy, the great sea channel to Rona before me, with stag picking through the seaweed on the beach only yards away. His writing inspired me yet again, and drove me to greater efforts to understand the land and the creatures around me.

We tend to be suspicious of eulogising, of enthusiasm, of the need to communicate a passion for a place. I can speak only from a personal perspective, with my feelings for Sand Bay – the location of the bothy – and the village itself summed up neatly on my final day there. As I drove over the mountain pass, I stopped for one final time, looking back down at the bay and the village. It was a somewhat gloomy day, with the first showers of winter sweeping up the channel, however the sun was fighting a heroic rearguard action and appeared occasionally through the gaps in the dark clouds. The bay was cupped under a large rainbow, looking like something out of a children's book, with the end of the rainbow seeming to emanate from the village itself. It made me smile as I climbed back into the car to head back to my life down south, the thought that – on so many levels – Applecross was the place where my rainbow hit the ground.

1

Arrival

The drive would take fifteen hours, a slow meander up virtually the entire length of Britain. Not having the first idea of what was required for my six-month sojourn as a crofter on the west coast of Scotland, I had opted to take along pretty much everything I owned. The vehicle to transport this paraphernalia was a shiny new Land Rover, now partially concealed by ropes, tarpaulins, pots, pans and the sort of kit I had guessed would be required for my protracted visit to the wild west. I had left behind my home in Bristol, and my girlfriend Antje, for what would surely be the experience of a lifetime.

Accompanying me on this great adventure was my new and gigantic dog Reuben. Ten days previously, Reuben had been living in a small concrete pen at the local animal rescue centre, a place overwhelmed with waifs and strays, and run on a shoestring by a group of hardworking volunteers. I had chatted to them about the project in Scotland, and they had recommended Reuben as the ideal companion – an easy-going, affable, get-along-with-anyone dog. He was all of these things when I finally met him. What they neglected to mention was that he was also the size of a Bengal tiger.

Each stop for petrol saw the accents change, the scenery subtly alter and the sun track slowly across the sky. Although this drive could be said to be the beginning of my adventure, the real seed had been sown for me many years previously in the magical pages of Gavin Maxwell's *Ring of Bright Water*. His story of raising an otter in the clear waters of the bay at Camusfearna bewitched me, much as it did an entire generation. His quest to escape the madding crowd touched something basic in me, even as a child, and for months afterwards I dreamt only of silent bays, rushing falls and dark hills. This obsession with the Highlands and islands never left me, remaining a distant aspiration through the days of my adolescence, then an indulgent pleasure as my career with the Royal Marines took me to Scotland for three years, where I explored the region on long pathless rambles through many a shadowy glen and sugar-white beach. Now the wanderings of my life, both professionally and geographically, had brought me to a point where I could follow my dream of establishing a crofting lifestyle.

My reasons for heading to the west coast were, however, very different to Gavin Maxwell's. He had a profound connection with the Highlands throughout his life, and sought the peace of the west coast after his service in the Second World War. He rejoiced in the solitude of the region, and of course in the animals around him. He had, however, a rare ability to establish wildly impractical schemes, and he was dogged by bad luck. He seemed to me to be a profoundly troubled man. Despite this darker side to his personality, he produced truly beautiful prose, writing what was a gentle poem to the region and the animals that had at last given him a measure of peace. As one of the finest wildlife authors of his generation, he left a legacy that outlived him, with books born of personal melancholy and an intuitive empathy with the natural world.

I, on the other hand, was very much a product of the system. After attending a minor public school – toddling up the drive at the age of 7 and re-emerging blinking owlishly from the front gate as a foppish 18-year-old – I had travelled for a few years before joining the Royal Marines. Showing unusually uncanny judgement, I chose for my military career the only eight years in the entire 300-year history of the Marines when they didn't go to war. My time was spent running around dressed as a bush, creeping up on people in training areas in Wales, shouting 'bang' (as we were rather short of blank training ammunition) and playing a great deal of rugby (very badly). Having decided that we were unlikely to be invaded by Welshmen who didn't like people with shrubs stuffed into their shirts who shouted at them, I eventually decided that enough was enough and left the forces to pursue a career in expeditions and marine biology. Almost immediately afterwards a global outbreak of small wars and terrorist incidents saw the Royal Marines swing into action relentlessly over the next decade, covering themselves in glory whilst I studied limpets in Plymouth Sound.

Having subsequently embarked on various expeditions to far-flung corners of the world, I was soon writing for travel magazines as well as fronting documentaries for a variety of satellite television channels, most of them watched only by my immediate relatives. Finally, however, the peculiar currents and eddies of life, through which we all bob helplessly, had driven me to this point – driving a Land Rover as far north as possible on the British mainland, surrounded by an assortment of kit and accompanied by a donkey-sized dog, towards a coastline I considered the most beautiful and mysterious on Earth.

Making your dreams come true is of course a very dangerous thing to do – some are best left as aspirations, a vision to sustain you as the rain drums on the office window. Moving to a remote cottage on the coast would throw up all manner of challenges that I may or may not be capable of dealing with – my dream could swiftly become a nightmare. A good parallel might be dating Naomi Campbell – outstanding in theory, but in reality there is the distinct chance of injury, and possibly even death.

But my stay on the west coast would at last answer the questions that had always dogged me about the crofter's lifestyle. Would it really be an idyllic life in the heart of a beautiful and peaceful coastline? Or would it be a nightmare of relentless toil, hostile locals, midges feasting on my pale English flesh, and ultimately a slow whisky-soaked descent into madness?

After an interminable drive, I finally pulled up, hollow-eyed and covered in flapjack crumbs, at an old school house in the tiny settlement of Callakille, perched at the water's edge on the mainland opposite the northern tip of the Isle of Skye. This quaint old building was to be my base for the next few days as my quest for the perfect cottage took place.

I had a number of possible bothies arranged for inspection over the next week, although this was far from my mind as I stumbled around the car to release the dog, who had been staring at me, catatonic with boredom, since Carlisle. The whisper of the waves and the crisp night air were the only hints of what was to come, with the oppressive darkness of genuine wilderness meaning that I fumbled my way to the front door of the house, mummy-like with arms outstretched, emitting the occasional low moan for effect. The key was tucked away under the doormat as promised, and the dog and I staggered into the warmth of the interior. Returning to the car to retrieve a few essential items of kit, I finally made my weary way upstairs, and pushed open the bedroom door. Dropping my bag, I crashed onto the bed like a dynamited lighthouse.

Ten blissful hours later, I was awoken by sunlight streaming through a gap in the alarmingly floral curtains. I sat up groggily, my senses hastily rebooting as I took in my new surroundings. Once I had figured out where I was, who I was, and why there appeared to be a wolf sleeping on my bedroom floor, I swung my legs out of bed and ambled onto the landing. Reuben followed me out of the room, yawning and stretching before giving a good-natured wag of the tail to indicate that all was forgiven from the day before. We descended the stairs to the kitchen, where I shuffled across to the kettle whilst Reuben slumped at my feet to begin the process of the morning nibble and scratch. Moments later, with one of us nursing a cup of tea and the other temporarily itch free, we made our way to the kitchen door. I threw it open and there we stood, stock still, frozen in sensory shock.

Reuben, whose entire universe until only days ago had consisted of a concrete pen and a small grassy exercise paddock, stared transfixed at the scene before him, ears pricked, almost vibrating with excitement. We were faced with a rolling field that led down to a rocky shore, the sea pulsing and sliding along a series of mini coves and inlets, the early morning sun sparkling on mini explosions of spray as each wave dashed itself against vertical rock walls. Across the sea channel – a rich, deep blue as opposed to the usual temperate green of British waters – stood the great mountains of the Isle of Skye, dusted with snow on their crenulated peaks. The clear morning air displayed the ridges of the mountains in serried ranks, creating an impression of a magical Tolkien-esque landscape beyond, at once forbidding and beckoning. Looming over the scene was a vast cornflower blue sky. It was excessively, ridiculously, riotously beautiful.

Reuben barged past me to race down the slope towards the sea, bounding and spinning in a display of exuberance that made me laugh out loud. Although not quite as demonstrative as the dog, my feelings were similar. Like a starving man ignoring a feast, I closed my eyes to the staggering view before me, and took a great breath of air, smelling the sea, the kelp and the damp peat – the essence of a personal Eden.

*

The Highlands and Islands of Scotland is one of the most sparsely populated regions in Europe. Although this has great appeal for the modern traveller, the reasons for such great tracts of wilderness have a distinctly dark side. Up until the Jacobite Rebellion of 1715, the Highlands were a series of self-sustaining communities divided into clans, a term derived from the Gaelic word 'clann', meaning family. Such was the bond of the clans with the mountains and glens that there were a dozen different words in the Gaelic language to describe a piece of high ground. Loyalties within the clan were strong, and centred on a chief, whose name was frequently adopted by those within the community. Such loyalty to both the name and the chief meant a high price for many during the rebellion and at the subsequent Battle of Culloden in 1746, in which the Highlanders suffered a devastating defeat at the hands of the English (and indeed their fellow Scots). This marked the beginning of a period of what can only be described in modern terms as ethnic cleansing, with entire populations forced out, the banning of Highland dress, and the use of the Gaelic language or playing of traditional music forbidden.

These communities consisted of small settlements, with the buildings created entirely from natural materials. With the forced dissipating of the clans, the settlements themselves were simply devoured by the landscape, returning to wood, field and stream over time, enfolded by the same materials that had given them birth. Today no traces of the old Highland communities remain.

Although this period is seen by many Scots as a time of oppression by the English, there was one other element that struck at the very heart of the honour of a famously fierce and proud people. Due partly to exorbitant rents demanded by the English, and also to fund extravagant lifestyles in England and southern Scotland, many chiefs – or landowners as they were now termed – began to turn their land over to more profitable sheep farming. The subsequent evictions of families leasing smallholdings, with many burnt out of house and home or violently relocated to the coast, were a betrayal of trust that has never been forgotten. On the coast the poor soil and harsh conditions meant great hardship, demanding the acquisition of new skills such as kelping – harvesting kelp for the alginates used to make glass and soap. Many crofters emigrated, many were forced to move to the towns and cities, and many died. The empty landscapes of the west coast of Scotland today echo with the memory of a lifestyle shattered by greed and betrayal, the wind whistling around crumbling ruins that stand like tombstones for families scattered to the four corners of the globe.

Given the less than glittering track record of the English in this region, it was with some trepidation that I traversed the coast road towards my first potential cottage. I was a middle-class Englishman, an ex-military man to boot, called Monty, wearing a waxed jacket, driving a Land Rover, with a black dog in the passenger seat. I might as well cycle slowly through central Baghdad, wearing a novelty George Bush mask playing 'The Star-Spangled Banner' repeatedly on a trombone. Of course I wasn't expecting to be dragged wriggling into a bush to be dispatched with a claymore, but in terms of displaying my origins I couldn't have been more blatant.

The first property I was due to view was described as 'run down', which in this part of the world could mean anything from an atmospheric bothy with views to die for, right through to a random pile of rocks in a swamp. The photo on the web was somewhat vague, and what's more had been taken in the lowest resolution possible. Squinting at the image from the other side of the room, a vaguely respectable looking bothy emerged, but plainly the only thing to do was to go and see it for myself.

The drive along the coast was glorious – a single track road writhing along a craggy shore with clear water beneath. Even Reuben looked up, the remains of the pig's ear I had used to lure him into the car sprinkled on his muzzle, to peer curiously through the glass. I opened the window to create a rush of clear air, Reuben's nose twitching beside me at the jet wash of new smells being fired up each nostril. Dogs of course rely on olfactory as much as visual signals, and I could only guess at what he made of the whiff of seagulls, seals, old seaweed and damp shingle that were barrelling through the window.

After a drive of over an hour, the first bothy hove into view. As I pulled up alongside, and before the engine had even stilled, I knew this was the wrong place for me. An air of complete desolation hung over the building, with sagging walls, hanging gutters, and floorboards rotted away to reveal dirt piled up by the winds that spun and whistled through the ruins. The building also had a tarmac road right alongside, an artery for an inevitable steady passage of tourists come the summer months. After the most cursory of inspections, I climbed back into the Land Rover and began to thumb through the map to find the next location.

This place looked more promising, an old chapel in a tiny cove on the Isle of Skye. If ever there was a mystical island in Britain, it is Skye, shrouded in legend and folklore, the favoured haunt of wandering poets and tormented writers. I felt a keen sense of anticipation as I crossed the bridge from the small town of Kyle of Lochalsh, the sea channel surging beneath me, with the shadowy blue mountains of the island rising ahead. A long drive up the coast led me to a tiny turning circle at the end of a rutted road, and here Reuben and I left the vehicle to search for the cove as described on the website.

It did not take us long to find the chapel, sitting with its back to the cliff wall, surrounded on three sides by steep rocks with the sea immediately ahead. Descending into a deep gulley that led to the cove, there was immediately a sense of preternatural hush, a stilling of the wind that magnified the echo of our footsteps as they crunched on the shingle. Reuben glanced up at me and edged closer, ears flat and tail low. This was a weirdly atmospheric place, and had I possessed large floppy ears and a bushy tail they would have been doing precisely the same.

Approaching the front door of the church, I leaned gently into the weathered wood, causing a theatrical creak on the hinges as it swung open. Reuben whimpered beside me, and seemed to be toying with the idea of jumping into my arms, Scooby Doo-style, with chattering teeth and trembling paws. It was as though the last congregation – obviously many decades ago – had simply closed their hymn books, stood from the pews and filed silently out of the church, closing the door behind them to let the church slowly turn to dust and fall gently in on itself. The pews remained in neat rows, the pulpit stood empty and silent at the front of the aisle, and light filtered eerily through the boards on the windows, specks of dust turning and glinting in the rays of a watery sun.

I walked slowly into the building, my feet crunching

on old glass and the wind rocking the door on its hinges behind me. The thought of spending six months here was not appealing. Actually, the idea of spending the next ten minutes here was rapidly losing its charm. I could have leased the place of course, but only if I fancied several months of ghostly apparitions, rattling windows, long-lost relatives paying me visits in the dead of night, and the prospect of emerging at the end as a white-haired, cackling maniac in a pair of soiled pants. By this stage, Reuben had turned tail and slipped out of the door, back to the Land Rover and the world of pigs' ears and nice smells. He had the right idea: the chapel, I decided, would not do.

With the long drives and the two viewings, the day had mysteriously slipped by, something I was to grow used to over the next six months. I wearily climbed into the car alongside Reuben, and headed back up the road to a small pub I had seen en route. Having checked into a tiny room, so bijou that every time I stepped out of bed I had to clamber over the increasingly annoyed dog, I went down to the bar and considered my options. Rather naively, I had only arranged three possible places to view, so I was in real trouble if the final bothy turned out to be as wildly unsuitable as the first two. For this last one on my short list, a place called Applecross, there hadn't even been an image on the website. My hopes were not high. I lapsed that night into a troubled sleep, and dreamed of Antje and home.

*

To get to Applecross, I had to cross a high road pass called the Bealach na Ba – translated as the Pass of the Cattle. Even on the map, this road looked a handful, crushed between densely packed contour lines that created a dark smudge on the map, the road itself a yellow line squirming in their midst, punctuated with black arrows showing areas of extreme steepness. This is the highest road pass in Britain, reaching an impressive 2,053 feet (615.9 metres) before plunging in a series of roller-coaster bends down into the tiny settlement of the village itself.

On arrival at the foot of the mountains, still white with the snow on their peaks, the road looked even more daunting, snaking its way into the dark shadows of the gulley that in turn led to a massive saddle that has always been a natural route over the mountains for crofters driving their cattle south. The mountain road had only been built in 1822; before that it was simply a path created by nervous cattle driven by sinewy drovers who averaged ten miles a day, a feat of considerable endurance as the winds howled and the cattle skidded and slipped over rocks shattered by the winter frosts. These drovers were legendary hard men, and were the only profession in the Highlands allowed to carry arms after the rebellion, in order to fend off predators and bandits. They were in even greater danger on the way home, returning along the same routes with their payment for the drive. They swiftly learnt to hide their wealth, with many melting down their silver coin and turning it into dull buttons on their ragged jackets. The great rebel Rob Roy was a drover, and their reputation persists to this day.

Even after the road over the pass was built, the winter snows made it impassable for half of the year, meaning that Applecross was in effect an island community, relying solely on stores and communications from the sea. Even the addition of a coast road in 1976 did not truly open access to the outside world, still requiring a singularly determined traveller to reach the village, which of course was precisely why I was so keen to visit the place myself.

The Land Rover roared and rattled up the hairpin bends of the pass, my knuckles white on the steering wheel as the land fell away beneath us. In a country where jaw-dropping views are standard, the view from the lay-by at the top raised the bar to a new level, with Skye sprawled in a sparkling sea in the distance, all curling bays and sharp ridges. The run down the other side to Applecross was exhilarating stuff, with my feet pistoning the brake and clutch, and Reuben being bounced from one side of the cab to another.

Applecross eventually hove into view beneath us, a scattering of houses in a broad sweep of bay, fringed by green fields and golden sand cupping the Atlantic rollers that whispered up the shallow beach. Applecross was also known as a'Chomraich in the Gaelic tongue – a place of sanctuary. This was due to a singularly determined Irish monk called Maelrubha, who in AD 673 set up a monastery and declared the land around it sacred. The monastery is long gone, with a church now standing on the site. The main village consists of a row of white houses on the shoreline, clustered very sensibly around a pub. The whole community nestles beneath the hills that rise immediately behind, with pine forests coating the crags and gorges, and the River Crossan murmuring over rocks and rapids before disgorging into the bay. Standing just back from the shore is Applecross House, an imposing white building built in 1740 by the Mackenzies – the main landowners of the region. It still stands four-square facing the sea, the white walls contrasting with the emerald green of the fields surrounding it. Dotting the fields were the gigantic red figures of Highland cattle, several of whom glanced up as we passed, peering at the Land Rover through shaggy fringes of trailing hair.

Leaving the village behind me, I travelled several miles along the coast road to reach my destination, the encouragingly named Sand Bay. The road rose gently over the brow of a rolling hill, and there before me was the bothy.

Dominating the scene was the bay itself, a huge spread of golden sand criss-crossed by a venous network of tiny rivulets and streams, all running into a gentle, ankle-slapping surf that flopped onto the beach. Even as I stopped the car on the rise of the road, I could see oystercatchers bustling along the shoreline, looking neat and manicured in their black and white dinner jackets, vivid orange bills probing the sand before them. At the back of the beach was a huge sand dune, more akin to Namibia than Scotland. It was dotted with footprints speaking of lengthy climbs upward, alongside grooves that marked many a shrieking descent by exhilarated visitors.

The right-hand side of the bay was dominated by a craggy headland, with the bothy perched defiantly on its tip. I drove down to the small car park at the head of the road leading directly to the bay. The walk down to the beach crossed first field, then bog, then dune, before finally giving way to the sand itself. This array of new environments underfoot meant that Reuben covered this distance by revolving rapidly as he moved, springing and cavorting, before eventually charging to the edge of the surf, then sprinting back to – rather unexpectedly – bite me on the bottom. I took this to be a sign of approval of his potential new home.

In a dip behind the peninsula stood the ruin of the main house, for which the bothy had served as a barn. Although little more than a pile of stones, it still held the memory of former grandeur, a single wall standing proud of the bracken with a fireplace at its base, the very heart of any Highland home. Scuffing the floor with the toe of my boot revealed the traces of old partitions between rooms. I was plainly not the first person to be seduced by the beauty of Sand Bay. This house spoke of success and endeavour, of a lasting occupation over many decades. It was a slightly eerie place, as old ruins tend to be, making me feel like a trespasser, treading on the memories and dreams of its last occupants.

The bothy itself was a two-minute walk from the main ruin. From a distance it did not look encouraging, giving the impression that it had been subjected to a fairly sustained naval bombardment, with gaping holes in the roof and a general air of neglect. It was only on arriving at the front door – or at the space where a front door should have been – that I realised how lovingly the bothy had been built. Over 150 years earlier, rough hands had gently piled stone on stone, with the gaps carefully filled by smaller rocks. It wasn't just a few weeks work that had created those walls; it was generations of training and apprenticeships, of learning a trade handed from father to son to create something that would stand on the shore for many decades. The walls were plumb straight, each several feet thick, standing proud having stared into the teeth of a century of Atlantic storms. Behind the bothy were the animal pens, known as the fank – three neat squares with the walls perfectly intact, topped with gigantic flat rocks to maintain their integrity through the fury of winter gale and rain.

I went and sat on the headland itself, watching the clouds scud gently across the sky, the mountains beyond the flat calm of the sound, and the waves roll onto the beach. Some things don't require a huge amount of thought – I had four good walls, pens for stock, a setting that I could only have dreamed of, a great deal of work to do, and a very, very happy dog. I was sitting next to Beachcomber Cottage.

2

The Build

Acquiring the bothy, even for only a six-month period, was going to be no easy matter. One crofter memorably described his property as 'a small patch of land surrounded by regulations', an apt term for a lifestyle cocooned in historical enmity, ancient and modern legislation, and contemporary politics. The Applecross Estate Trust website had given me details of a local contact who could advise me on the legitimacy of my project, and I walked briskly back to the car park, phone held aloft, squinting at the screen for a signal. As I walked over the top of one of the larger dunes, my phone beeped into life, and I immediately dialled the number.

'Archie MacLellan, hello,' said a friendly Scottish voice.

'Archie, it's Monty Halls. I'm looking at the bothy down in Sand Bay, and would very much like to lease it off you for six months.'

'Aye well, it's probably best if you pop up and have a chat with Mike Summers in the Estate office, and then we can take it from there. Mike's really the man who can give you an idea of what is possible in terms of the renovation, and he's very much my contact on the ground out there. Just pop in and tell him what you'd like to do, and then we'll see.'

Archie's tone was friendly and open, and I felt a powerful surge of optimism rise within me. Perhaps this wasn't going to be the delicate negotiation I had initially envisaged.

Buoyed by this minor triumph, I positively skipped back to the Land Rover and bundled the dog into the back. As the Estate foreman for the Applecross Estate Trust, Mike would be a key figure in the acquisition of the cottage in Sand Bay, and I was keen to make an impression. Drawing up outside Applecross House, I found Mike's office tucked discreetly around the back, and tapped on the door.

'Come in,' said a low voice. Perhaps it was my fevered imagination, but I seemed to detect a level of reticence even in those two short words.

Mike stood from his desk as I walked in, a fit-looking man honed by years of hard physical work, with lines on his face etched by wind and weather. He was wearing stained blue overalls tucked into battered wellies, the one-size-fits-all attire of a man of the land. He extended one hand in a firm handshake, his palm rasping against mine like a piece of driftwood. His smile of welcome certainly seemed genuine, although his opening words caught me somewhat on the hop.

'So, you want to do up the old bothy, do you? You know that many of the bothies were destroyed in the clearances

– the technique was to burn the roofs so no one could get any wood to rebuild them. It was a savage thing to do, something that leaves scars that'll never heal. People round here have long memories, you know.'

Although I had expected to be questioned about my motives for taking over the bothy, I hadn't really anticipated being singled out and blamed for the Highland clearances before even being invited to take a seat. I mumbled something between a consolation and an apology, and stood politely awaiting the next development in the exchange. If it had started, 'And as for the Boer War', I wouldn't have been too surprised, but actually Mike waved me into a seat and looked at me intently, blue eyes boring directly into mine.

'The bothy in Sand Bay is actually a lovely old building. It looks like a ruin, but the stone work is some of the best I've seen on the Estate – if you look closely at it you'll see there is real craftsmanship in there. I'm keen for you to tell me what you're trying to do with it, and how you see it developing after you leave.'

I took a deep breath and over the next twenty minutes I explained my plans, mentioning that my DIY skills were at best sketchy, at worst potentially fatal. I also noted that my experience of raising livestock consisted of a single miniature pig that my girlfriend at the time had decided to rear, as it had been rejected by its mother at a local wildlife park. After a hugely entertaining three months we handed a healthy boar (called Phoebe for reasons that are far too complex to go into here) over to a local petting zoo.

Much of this was met with a stony silence and an unblinking gaze, Mike's impassive blue eyes showing no trace of emotion. Feeling the first stirrings of anger rise in me, I opted for a final gambit:

'Mike, I'm just a regular chap who has always wanted to do this. I will give it 100 per cent, I'll read up everything I can, consult locals about everything I need to do, and put my heart and soul into it. This won't fail through lack of effort, I promise you, and you'll be left with something worthwhile to hand on to the next tenant.'

With that, I trailed off and waited miserably for the verdict. By now Mike had steepled his fingers and was peering at me from over the top of cack-rimmed fingernails.

Eventually, he spoke: 'Well, let's give this a go, shall we? I'll help you out in clearing the place up, make sure your materials are delivered promptly, and offer advice when you need it. Make sure you ask lots of questions, use the knowledge around you, and who knows, you may even pull this off.'

It was a spectacular turnaround in the conversation, and I jumped up to thank Mike profusely. He smiled for the first time, then waved me away with a weathered hand, turning back to his cluttered desk.

With that, it appeared, the interview was over.

*

Having booked myself into a holiday cottage on the estate, I could at last sit down with a coffee and consider the task that lay ahead of me in terms of the renovation. My relatively miniscule budget meant that I had to be in the bothy within a couple of weeks, simply not being able to afford a long stay in any form of holiday accommodation. As the building at present consisted of four walls – albeit very nice ones – topped by some rusty tin that overlooked some crumbling animal pens, I needed to get cracking. I wanted at least a modicum of comfort, not relishing the prospect of shivering away the nights crouched over a sputtering candle flame or curled in a foetal position in a damp corner, coated in midges and pawing feverishly at my flesh. Maddened by caffeine, I resolved that the bothy would be habitable within a fortnight.

The only way to do this was to throw manpower at the problem. In this respect, I had arrived in Applecross at precisely the right time of year. April sees the village just beginning to stir from the state of mild hibernation required to survive the long winter months. From November onwards, the hatches are battened down, fires are stoked, and life revolves around the pub. When the bluebells carpet the ground within the woodlands and tiny lambs begin to appear, so the village begins to fill with seasonal labour drawn in from the surrounding countryside, doors are thrown open, and there is a general air of greeting a new summer. Spring is always a time of optimism and renewed vigour anywhere in Britain, but here in the land of the brutal, savage winter, it is met with a true sense of rebirth, no doubt a throwback to the days when the very survival of the residents could be at stake.

The obvious source of manpower was the pub, the famous Applecross Inn. I had heard of the pub before hearing of the village, the guidebook telling me that this was the source of good nights, great food and hospitality on an epic scale.

It was a short walk to the pub, the lights glowing in the dusk and the babble of voices growing louder as I approached. As a newcomer to the village, I was rather nervous as I pushed the door ajar – anticipating a darts-stopping-in-mid-air silence as I entered. But within moments I was nursing a beer, warming my hands on a spitting fire, and chatting to someone about Reuben. The dog turned out to be the ultimate device for removing social barriers – it's tricky to ignore someone when your gigantic hound has just nicked one of their chips and is now attempting to sniff their crotch. In addition, it was a huge help that Reuben was also a hopeless tart, gazing with doe-eyed adoration at anyone who scratched behind one of his ears or rubbed his belly. Many of the people in the pub were locals who worked their own crofts, and dogs were a crucial part of their lives. They could spot a good – if somewhat ill-disciplined – dog a mile off, and soon Reuben was working the room like a seasoned politician.

One of the first people to comment on Reuben was a lean-looking man at the corner of the bar. He had the look of a traveller, casually dressed in faded clothes, with dense ginger hair and magnificent sideburns running amok down weathered cheeks. Smiling shyly, he introduced himself as Sam – patting Reuben and simultaneously shaking my hand.

'That's a lovely dog you have there,' he said, rubbing Reuben behind one of his ears, 'and just a big pup by the look of things.'

We swiftly got chatting, and Sam waved me towards the bar stool next to him. As you invariably seem to do in these circumstances, I asked Sam where he lived in the village. He smiled gently.

'Ach, you know, here and there.'

Sam was indeed a nomadic presence on the coast around

the village, moving from job to job, occasionally living rough or spending periods in rented accommodation. Although in modern society we tend to mistrust anyone who turns their back on stability and the vast array of mod cons that make our life easier, Sam was a throwback to another age on the west coast. Before the clearances, nomadic workers were welcomed not only as a vital source of labour in times of plenty, but also as the source of news from far afield. Even after the clearances, there were groups known as 'the Summer Walkers' who would roam the countryside, too proud to settle in a single location. With them travelled the oral traditions and a near mythical understanding of fieldcraft. They would be warmly welcomed into households, where food and lodgings would be provided as families sat mesmerised, catching up on events in distant communities. The arrival of such travellers was a source of celebration, a harbinger of good times and a link to far-off relatives and friends. As we talked, it dawned on me that Sam's quiet demeanour, dry wit and obvious intelligence would have made him perfect for this role, making me feel that he had been somewhat robbed by an accident of time.

As we spoke, Sam told me a little about the village, about how the busy times were just about to begin after the lean period of the winter months. He gestured around the pub as he spoke, pointing out the locals and engineering introductions as people made their way to the bar. He was plainly regarded with great affection by all, a will o' the wisp wanderer who had settled briefly in the community – a scene echoed throughout history in the Highlands.

After several pints, I asked Sam if he would be willing to help out with some work on the bothy.

Again that gentle, enigmatic smile. 'Aye, if you like. You seem a decent enough chap.'

It was the start of one of my great friendships in Applecross. The combination of Sam's contacts in the community of Applecross and Reuben's preponderance for other people's chips meant that, when I wobbled my way out of the pub several hours later, I had accumulated the beginnings of a small workforce, their numbers scrawled on a damp beer mat pressed soggily into my pocket. Work would begin the following day – the start of the transformation and rebirth of the bothy.

The next morning, nursing a dull ache behind the eyes, I made sure I was at the bothy first thing and was dismayed to see Mike from the Estate office already on the roof removing the old tin. I was slightly unsettled that he was there before me – I had wanted to prove my mettle by being first on site. But regardless of Mike's early appearance, I knew it was going to be a rough start: the prospect of vigorously battering a rusty tin roof with a lump hammer in the glare of the sun made me quail, a total sensory assault for a man nursing his first west coast hangover.

Mike noted my arrival with the curtest of nods, and turned back to the work. I couldn't help noticing that he wasn't wearing gloves, and already his hands were a latticework of cuts and grazes, weeping blood down his fingers in tiny rivulets. I tucked my own gloves away, bought in B&Q before I drove up to Scotland and, for some reason best known to myself, bright pink. Hopping up onto the roof, I asked Mike if there were any other tools I could use.

Glancing up in surprise, he indicated a hammer and crowbar with a jerk of the head. 'Aye, over there.' With a quick disbelieving glance at the gloves hanging out of my belt he went back to work.

My hammer was tiny, and I quickly began to feel that I was using one of those squeaky mallets you give a toddler. My crowbar also seemed to have been sourced in Toys R Us, and would have been excellent for prising apart Lego, but wasn't the best on a hundred-year-old roof built to withstand the worst of the west coast weather. Nonetheless I attacked the roof with gusto, glancing reproachfully at Mike as he frisbeed razor-sharp tin past my shins and onto the floor.

We quickly stripped away the tin sheets of old roof, leaving the bothy looking painfully exposed. Further work the next morning saw the old timbers of the roof frame removed, until finally only the ancient stone of the walls remained. This was as raw as the dwelling would ever be, the base from which my dream croft would hopefully rise.

*

It is probably worth pointing out that I am about as far from a DIY specialist as it is possible to be. A visit to B&Q leaves me feeling completely overwhelmed, with great multicoloured shelves holding endless variations, shapes and sizes of the item I've gone in for. What should be a straightforward search for something as simple as a hinge becomes a tortuous experience of overloaded senses, and I always end up asking some smirking, speckled youth for help. This invariably results in a snort of derision and comments along the lines of 'Nah, mate, you want the laminated crenulated triple-backed version, what you're holding is the indoor grade for reinforcing kittens' boxes and making doll's houses.' This leaves me feeling badly unmanned.

Still, if I was going to renovate the bothy, I needed some kit. When I asked Mike where would be the best place to buy it, he went somewhat dreamy-eyed and stared into the middle distance, murmuring reverentially, 'Highland Industrial Supplies, of course.'

Highland Industrial Supplies – or HIS as it is appropriately known – is something of an institution in Scotland. It's the bothy-builder's B&Q, a Homebase for the horny-palmed. Having travelled to Inverness, I entered the car park outside HIS and stopped the Land Rover next to a burger van that was serving enormous bacon sandwiches to a patient queue of flat-capped, ruddy-cheeked farmers. On entering the shop, I saw gigantic bits of machinery bolted to the walls, with rows of industrial-sized wheelbarrows and sit-on mowers the size of Chieftain tanks lined up on the floor.

Spotting my disquiet, an assistant came bustling over and asked me what I was after.

'I'm . . . er . . . rebuilding a bothy.'

To his credit he barely flinched. 'And what tools do you have at present, sir?'

'None whatsoever.'

This chap plainly enjoyed a challenge, and wordlessly led me to one of the larger barrows. Placing me behind it, he began to take me round the shop floor, chucking gigantic bags of nails and some very exciting-looking tools into it, our progress punctuated by a series of clangs and bongs as tools met barrow. When we got to the counter, the barrow was creaking and my knees were buckling, but it seemed to me that we had the beginnings of a set of gear with which to lay about the croft.

As I packed the kit away and prepared to pay, I noticed an interesting phenomenon. As the farmers around me paid for their equipment ('Aye, that'll be twelve supersize crowbars, seventeen hinges big enough to hold a barn door closed in a howling storm, some cattle wrenches and a pig clencher'), they all added something rather surprising on the end of the list.

'Oh, and ten bottles of Skin So Soft, if you will.'

These sat incongruously behind the counter in neat rows of sky-blue bottles, every inch the moisturiser of choice for ladies who lunch. This raised the somewhat bizarre image of massive swarthy hill farmers rubbing in Skin So Soft before heading off to work, retiring many years later with complexions that a supermodel would kill for. Noticing my quizzical expression, the assistant laughed.

'You'll nae find a better repellent for the midges in high summer. If I were you, young man, I'd get plenty of it where you're going to be living.'

I certainly needed protection from what I had been assured was the scourge of the west coast and, if it worked for them, then I wasn't about to ignore what seemed sage advice. I purchased several bottles, dropping them into the barrow to complete my HIS shopping experience, and headed back to the bothy, ready for work.

*

My plans for the bothy were very simple indeed. The first priority was to put a roof on – I was assured by all and sundry that relentless sunshine and shimmering heat were not summer-long phenomena on the west coast. Next on the list was an extension, the middle-class boy in me wanting to watch the world go by whilst sitting on a sofa bathed in the gentle glow of the evening sun. I also needed gates for the animal pens, and somewhere to put my vast pile of new and embarrassingly shiny tools. The aim was to accomplish all of this in the first few days, before moving on to the serious work such as the deer fence, water supply and electricity in the weeks that followed.

Happily, Sam's hastily recruited workforce proved to be both enthusiastic and capable. First was 'Kayak' Mike, a shaven-headed, fit 42-year-old Scot. Although his main role in life was as the local kayaking and climbing guide, he was also a dab hand at hammering together the old roofing timbers, creating passable imitations of gates for the impending arrival of the stock. He also shared my services background, and the two of us would while away many an afternoon telling stories of questionable accuracy about our time in the military.

Andy was an ex-biker from Yorkshire who had holidayed in Applecross several years previously, returning with his family to take root, happy in the knowledge that he had found the place where he wanted to spend the rest of his days. He was a hugely capable joiner, wielding two nail guns with gusto, wooden-framed structures magically appearing around him. Alicia was a petite English girl who had also travelled to Applecross to work as an outdoor instructor four years previously, never to leave – something of a theme in the village it seemed. She applied herself with limitless energy to various odd jobs around the build, one moment wielding a gigantic crowbar, the next clearing several decades of dense brambles from within the fank, only the top of her head showing as she ploughed her way steadily to the far gate.

For a fortnight, the bothy became a hive of activity, with long lunch breaks to savour the spring sunshine. As the work progressed, I swiftly discovered that there are very few problems in DIY (and possibly life) that can't be solved with a large mallet, a bag of ten-centimetre nails and some swearing. The new roof was up in a couple of days, meaning that I could turn my attention to building the extension. This wasn't simply to sit and stare at the view, although that would have been reason enough; it was also to act as a hide to film and photograph the wildlife that would scull through the channel over the course of the summer. I positively attacked the project, moving slowly and inexorably from the bothy to the seashore, the extension springing up in my wake. I even put in a vast window, and it was only when I sat before it for the first time, the sand stretching away before me and the sea glittering beneath the mountains, that suddenly the expression 'bay window' made perfect sense.

Sam had decided at an early stage that he should tackle the considerable problem of vehicle access to the bothy. When we started, the only route in was a twisting track through the dunes, ending in a swampy patch that in turn led to a ditch with a stream running at its base. The early trips to the bothy in the Land Rover were terrific fun, and I found a number of tissue-thin excuses for charging up the old track, the engine roaring and the steering wheel bucking in my hands. This all came to an abrupt end after a bout of heavy rain when, in a moment of particularly exuberant and poorly judged driving, I veered onto the edge of the ditch and immediately bogged myself in to the wheel arches. I was towed out by a passing tourist in – and this really was the height of humiliation – a very shiny new four-wheel drive, which had plainly up to that point been used only to take Lavinia and Tarquin to their prep school.

This would not do at all, and Sam duly took on the task of creating a bridge. He was simply unparalleled when it came to moving huge rocks around, a tectonic force powered only by a couple of rolled ciggies and coffee of hallucinogenic intensity. He set to work immediately, vanishing into the middle distance to appear moments later carrying a fridge-sized rock which he frisbeed into the ditch. This was repeated steadily throughout the afternoon, until the ditch was neatly spanned by huge rocks, with the stream chuckling away merrily beneath. It was a bridge that would still be there in a thousand years, and Sam at last stepped back, dusting his hands on his tattered trousers.

'There we go, Monty,' he said, with that half-smile of his. 'Le Pont du Sam.'

A few days later, it was Kayak Mike's turn to impress. A fine carpenter as well as an affable presence around the build, he could turn his hand to most things and, once he had built the gates, he turned his attention to the substantial pile of old roof timbers. After studying them for a moment, he set off determinedly for the small courtyard at the back of the bothy. There he found me shovelling soil into the raised vegetable beds.

'Monty, I've had an idea. How would you like a shed?'

'I would like a shed very much indeed.'

'Then a shed you shall have. How about over there?' This

was accompanied by a vague wave at the side of the bothy.
'Perfect.'

'Then stand back and watch the master at work.'

Rolling up his sleeves, he squared his shoulders at the pile of old timbers and duly created a substantial lean-to in a single day, complete with neat internal hanging spaces and shelving. The bothy seemed to be miraculously emerging around me, a time-lapse series of images of the ruin becoming whole again. Perhaps more importantly, as we laboured in the sunshine on our communal mission of bringing life to the bothy, sipping coffee as strong as molten metal and sharing the stories of our own lives that had brought us to this point, I was acquiring several new friends.

*

Flushed with success after his work on the shed, Kayak Mike invited me to the pub for a quiet pint as the day, and our first fortnight at the bothy, drew to a close.

This was my first opportunity to meet Mike's wife Linda, an important figure in Applecross, as she was very much the social hub of the village. Linda was a local, born and bred, and, although I had heard her name mentioned, I had no idea what she looked like. Mike – as the local kayak instructor and mountain guide – was every inch the outdoor man, fit and tanned with an economy of movement, word and deed that spoke of many years in the hills. He cared passionately about the west coast and, while we waited for Linda to arrive, he told me at length about how the region was historically mired in the politics and feudal by-laws of yesteryear, frequently preventing the development of small local businesses and stunting economic growth. Nonetheless he talked optimistically of the future, of his plans, and of the real chance for transition as local communities took responsibility for their own futures.

'It's an exciting time, Monty,' he said, eyes shining with enthusiasm 'The region has such potential, something we're all just beginning to truly appreciate.'

As we drank our pints at the bar, I was surprised to see what appeared to be a supermodel enter the pub. Blonde and impossibly glamorous, she was impeccably dressed in knee-length white leather boots, tight white shorts, and a halter top that showed off tanned shoulders over which a curtain of blonde hair swished as she turned her head and smiled in our direction.

Mike looked up and waved her over. 'Monty, this is my wife, Linda.'

She smiled the friendliest of smiles, and extended an immaculately manicured hand. 'Hello, Mr Superhuman.'

This caught me somewhat unawares and, seeing my palpable shock, she laughed delightedly and turned to Mike to plant a quick kiss on his cheek before speaking again to me. 'I googled you – the truth is out.'

This was something of a guilty secret for me, and something I had rather hoped to keep close to my chest whilst in Applecross. Many years earlier, I had entered a televised competition attempting to find the ultimate all-round performer in the country. This wasn't quite as prestigious as it sounds, as the real players – the Steve Redgraves and Sally Gunnels of this world – had declined to take part. A relatively rag-tag bunch of competitors from various disciplines turned up on day one: climbers, jet-ski champions, explorers and entrepreneurs. I was there simply as an ex-Royal Marine, and rolled up with nothing more in mind than thoroughly enjoying the experience.

Over the next few months, we were probed, pushed, and tested to our absolute limits in a set of exercises devised by leading experts in human performance. I found some useful things out about myself (I have, for example, poor spatial awareness), and some not-so-useful things ('Sir, you are the loudest vomiter in the history of this facility,' to quote an instructor at the G-force machine in Farnborough). I was declared the winner at the end of a year of filming – which was nice – although I'm still quietly convinced it was all rigged. Since then I had endured years of abuse from my friends, mainly of the pants-outside-the-trousers variety. Even in Applecross, it seemed, there would be no escape.

Linda and Mike were the friendliest of company, and the beer flowed freely as the hours slipped away. I was amazed as the evening rolled on just how many of the locals I had got to know over the previous weeks, with many a nod and a smile exchanged as the door swung open and the pub gradually filled.

Late at night, when the background babble and laughter made it necessary to lean forward to hear what Linda and Mike were saying, I finally plucked up the courage to ask the question that had been bothering me all night.

'Chaps, I hate to be rude, but how did you meet? You seem quite . . . different.'

Mike smiled, and Linda took up the story, Armani sunglasses perched on top of her head.

'Mike was here fifteen years ago on an army exercise

when I was waiting on tables. I was only 14, so in the finest teenage tradition I sent my friend over to tell him I fancied him. He told me to come back when I was older. So I waited and waited – and seven years later I asked him again. We got together when I was in my early twenties, and were married soon afterwards. He is, essentially, my bit of rough – and every girl secretly wants a bit of rough.'

The evening rolled on, ending up in Mike and Linda's house where substantial drams were issued, guitars played badly (me) and brilliantly (Mike). A number of appalling jokes were also exchanged, with perhaps the finest coming from Linda as she picked up two prawns and sat them on top of one another ('Prawnographic'). I awoke the next morning on the sofa, one cheek resting in what appeared to be an old sandwich. I made a mental note never to try to match Mike in guitar playing, or Linda in wine drinking, again – advice I would habitually ignore over the next few months.

I decided to sneak away to avoid disturbing them as they slept, walking through the empty streets of the village in the early morning. The low sun bathed the pines lining the road with a soft light, daffodils punctuating the forest floor with starbursts of a brilliant yellow. The bay was flat calm, the surface ruffled only by the sedate passage of a group of greylag geese, early arrivals from their sulphurous winter home in Iceland. The cool air drifting off the water revived my senses, making my pace light and my stride long as I walked the coast path to Sand Bay.

Finally the bothy came into view, sun glinting off the new roof, the extension looking somewhat incongruous springing from the ancient stonework. As I walked down the hill by the dune, my pace quickened, carried down the hill by the irresistible forces of gravity and the intoxicating sight of my new home.

*

As the bothy now had sound walls, a door in place and a splendid new roof, it was time I moved in. On the day of the move, my first day as a resident on the west coast as opposed to a transient tourist, I awoke in my holiday cottage like a child on Christmas morning. I had packed the Land Rover the night before, and wasted no time in bundling the dog on board and rattling over to Sand Bay. Breasting the rise of the hill that overlooked the beach, I pulled the vehicle over to one side and studied the view before me, leaning forward so my chin rested on both hands as they gripped the top of the steering wheel.

The tide was at its lowest ebb, revealing the beach in all its glory, its golden acres spread before the bothy. A gentle surf rolled onto the beach, with two tiny rock islands appearing a few yards offshore, marked by small crackling wavelets around their base. The main burn from the hills behind the beach glittered and writhed its way across the flat sand, widening and slowing as it hit the surf line, vanishing into the immensity of the sea. At the head of the beach was a strand line of dark kelp, marking the highest point of the tide beneath the dunes and lush slopes of bracken and heather. The great dune dominated the slope down to the beach, with a contorted rowan tree at its summit, sculpted by a thousand storms into a bent, twisted set of limbs that carried the memory of dark nights when the winds shrieked and the rain hissed. The bothy itself sat rather self-consciously on the headland, resplendent in its new roof.

I drove down to the headland, rocking, revving and slewing through marsh and over small ditches with chuckling streams in their base. Pushing open the new door, I surveyed what would be my home for the next six months. The bare walls looked beautiful, arrow straight with the courses of stone carefully layered to create neat interior lines. The roof created an inner sanctum that offered security and comfort from the outside world, somewhere to return and shake out soaking coats before warming hands by the fire.

I unrolled my sleeping mat, arranged my bag on top, and made a neat pile of my wash kit, placing my cooking gear and camping stove alongside. I lay down on my back, hands linked on my chest, and stared at the new beams of the ceiling as the sunlight shone through the skylights, creating two shining pillars that played on the rough stone of the floor. Reuben sniffed at each dusty corner of the interior before slinking over to join me, circling twice before slumping down and sighing in contentment. I closed my eyes and listened to the gentle thump of the surf, the distant call of the oystercatchers, and the breeze as it whispered around the craggy walls, a chorus of welcome from the beach and sea for the new resident of the cottage after a century of silence.

3

Stocking up

The build took several more weeks to complete, a period that by happy coincidence saw some of the best weather Applecross had experienced for a decade. For day after day the sun arced across a cloudless sky, and we all quickly developed mahogany tans. Plants and flowers that had cautiously emerged after the hard frosts of winter exploded into riotous life, with bluebells carpeting the forest floor amid grass of such emerald intensity it seemed to glow with a life of its own. The air hummed with insects drugged by the heat, and the mountains in the distance shimmered and pulsed as though under the ripples of a racing stream. Reuben gasped and huffed, a black dog trapped in a dense overcoat. He would seek out whatever shade was available, lying full length and panting heavily. Eventually he would abandon the land and take refuge in the sea, gulls scattering as he raced into the shallows.

Living in the bothy meant that my relationship with the surrounding land and the sea became much more intense. When the workforce departed, I was completely alone on the headland, the beach stretching away before me, the sea rustling and whispering into the middle distance. This led my eye inexorably to the islands of Raasay and Rona, with the great mountains beyond a massive, primeval presence on the horizon. But I quickly discovered that I was not the only resident in Sand Bay, with life bustling around me throughout the day.

The beach changed not only from day to day, but from moment to moment. The gentle slope of the sand meant that the tide raced towards the land, bringing with it new life and driving clamorous hordes of sea birds at its front. Oystercatchers whirred low over the tops of the waves, contouring the hills and valleys of the surface with their wing tips millimetres from the crackling wavelets, at once part of the air and yet bound to the water beneath. They would settle in gaggles, probing the sand with bright orange bills, always at the edge of the surf, high-stepping ahead of the slithering foam of the low waves.

Slightly higher up the beach, the gulls were an uncouth mob, squabbling over some scrap within the seaweed, or sitting plumply on the sand with their chins resting on the beautiful eggshell grey of their puffed-out chests. The herring gulls always seemed a harmless presence, rather like a crowd of bored children loitering in a shopping mall. The greater black-backed gulls were a different matter altogether, a sinister presence pacing the shoreline or patrolling offshore, heads turning slowly, seeking out the weak or the unwary. Close up, they were massive and powerful, with a baleful glint in the shining garnets of their eyes.

Just offshore, in the deeper water of the channel, grey seals would regularly appear, their long noses in comical profile as they surveyed first the beach and then the tiny cottage with its new occupant. One seal in particular seemed fascinated by my presence, and would scull to within yards of the rocks, staring directly at me as I moved around the area of the bothy, following my movements precisely. This fascinated Reuben, who would stare with his head on one side, bewildered at this strange dog that had appeared from beneath the waves. He would slowly move towards the seal, freezing every time it looked at him, like a child playing statues. When he got within a certain distance, the seal would dive in an explosion of foam, to reappear fifty metres further out into the channel, where once again its slow progress towards shore would begin. Reuben would wag his tail, glance back at me, then sit and wait for the cycle to begin again.

As the tide raced up the gentle slope of the bay, great swathes of the beach would be devoured, until only a sliver remained, a waning moon of white sand. The water would now be within metres of the bothy, crackling and sluicing at the rocks at the bottom of the stone slip. The bay itself would be transformed into a glassy cove, shining and iridescent at the base of the great dune and the deep green of the bracken-clad hills. The bird life would now be condensed in a narrow band of seaweed and flotsam, rich pickings where the land met the sea, with a jumble of sticks, leaves, and kelp providing a matrix of scuttling, crawling life with a shrieking, squabbling mass of birds overhead.

The land around me had, to a degree, been shaped by

two forces – the elements and the crofting that had taken place on it for centuries, with man and nature vying in an ongoing battle for ownership of each acre. The low ground around the bothy was coated in bracken, unfurling in the sunshine to greet the new summer. As it was still late spring the croziers – the delicate green heads of the bracken shaped in tight whorls like the head of a violin – were gradually unravelling to reveal the fronds within. Soon the hills behind the beach were a twitching, rustling carpet, into which Reuben would vanish for minutes on end, his passage marked by the waving periscope of his tail and line of thrashing undergrowth ahead.

Bracken has featured strongly in the history of crofting, being used as bedding for animals and a pest repellent in the byres. The latter may well be due to the young croziers possessing a series of knobbly warts known as nectaries, which secrete a sweet sap that attracts ants. They in turn protect the crozier against marauding insects – one of those little miracles of cooperation between plant and animal. Bracken is also highly carcinogenic, meaning that a great many domestic animals that go to slaughter on the west coast have gut cancer. Most wild animals avoid feeding on the bracken, and so the plant continues to prosper, fiendishly well equipped to cope with all comers.

By contrast, the moorland of the higher ground looked stark and dead, still emerging from the trauma of winter. In May, very small yellow tormentil flowers began to appear, the name deriving from the medicinal qualities they were once thought to provide against diarrhoea and other torments of the stomach. It was on the rocky ground that nature had really run riot, with nodding groups of primroses appearing in bursts between the rocks. Sam assured me that these were here in numbers not seen for many years, a measure of a mild winter and an early spring, nature's alarm call as the climate slowly changes at the hand of man.

The salt marsh in the low ground leading to the bothy was a spongy quagmire, and seemed to me to be nothing more than a dense mat of coarse grass. On closer examination, I could make out patches of bog moss, a very important plant indeed, as it decays very slowly to create peat, the fuel of the crofter through the ages.

On my hands and knees in the salt marsh one morning, digging out the rear wheel of the Land Rover yet again, I saw a series of delicate, deep pink flowers. Muddy thumbing through my wildflower guide moments later revealed that these were sea thrift, which survive even when covered by the highest spring tides. I returned a week later when the marsh was under thirty centimetres of water at the peak of the tide, and saw the flowers defiantly swaying in the oily swell. A salt marsh is, it would seem, no place for the weak. Further muddy tramps around the marsh revealed sea arrowgrass, buckshorn plantain and sea milkwort – all little miracles of adaptation and resilience. Their dogged survival in the face of salt, spray, flood and storm was a stark reminder to me – I would have to develop my own abilities to adapt on this coast, or face an ignominious retreat to warmer climes.

*

The ruins of the bothy now had a smart new suit of modern materials, although the old walls beneath the new roof still remained the dominant feature. Visitors would reach out and touch the walls involuntarily, and I would often see their hands gently stroke the rough stone work as they chatted. It would always make me smile within, as though the history, craftsmanship and memories caught up in the old walls proved an irresistible sensory draw, the ancient bothy reminding any visitor that it was her who remained the very heart of the new building.

Atop the stonework, the smart new roof gleamed dully in the sun, made of grey metal sheets interrupted by two large skylights, harvesting the light and flooding the interior with soft colours. The old dipping pen at the back of the bothy had been cleared, everything within removed from the site due to the potential presence of organophosphates from the old sheep dip. This was a truly dreadful, insidious chemical, and stories abounded of the great strong farmers of yesteryear sitting at home today, trembling physical ruins, their bodies ravaged by the chemicals within the dip.

Having thoroughly cleaned out the dipping pen, soil from inside the old bothy had been placed within the raised walls. This was based on the sound principle that the bothy had been used as a barn for over a century, and the compacted dirt on the floor was essentially a layer cake of ancient dung.

In a corner of the small courtyard behind the bothy sat the composting toilet. I was inordinately proud of this structure, having built it from scratch. I had stood in the hole that Sam had dug and gradually hammered bits of wood together around me, ending up several hours later with my head sticking out where the loo seat would sit. Clambering out proved tricky, however having done so I continued to build as the dusk settled around me, and by the end of the day had a very passable outdoor dunny.

I had also encircled the entire bothy with a deer fence, partly to keep the deer out – I had been assured that they would ravage a vegetable patch within minutes – and also to keep stock in should they escape the fank. I had been told that native breeds of pigs and sheep were masters of the sneaky escape bid, and thought that double fencing would create a Stalag Luft-style escape-proof croft. Deer fencing is serious business, two metres high, and supported by posts the diameter of cabers. Having seen a red deer effortlessly hurdle a roadside fence over the pass, I could see why.

The extension proved to be a complete triumph, with a quick trip to Inverness turning up a hideous old fake leather sofa which I wrestled through the door. Sitting with my feet up when the wind blew, staring through the huge double window, it felt uncannily like being on the bridge of a ship, with the woodwork creaking, the windows rattling, and the sea crashing on the beach outside. It didn't take a huge leap of imagination to feel the extension roll and pitch, and imagine it breaking free to set sail into the dark waters of the sound.

At each corner of the bothy sat a large water butt, collecting over 800 litres of precious water when it rained. The beautiful sunshine, unbroken for weeks on end, meant that the butts remained dry, causing me to trek into the village several times a day with jerrycans, hefting them back into the Land Rover to rattle back to the bothy. This was a tiny glimpse of the kind of logistical problems faced by the crofters moved from the high grounds during the clearances, away from settlements sited next to rivers and streams and onto patches of barren coastal ground. Indeed every day presented me with a set of small challenges, a constant initiative test that meant I climbed into bed every evening physically and mentally spent. Perhaps the only way to truly measure yourself is when the modern tools of life are stripped away, when the problems being faced cannot be instantly solved with a credit card and a mobile phone. The bothy kept me constantly thinking, constantly solving and adapting, pushing me in ways I had not experienced for many years.

*

My next task was to plant a vegetable garden, ideally to be harvested during the end of the summer. I had romantic visions of crisp lettuce leaves, with dewdrops glinting on them like diamonds on green velvet, served with juicy beetroot, haemorrhaging purple juice onto potatoes the size of rugby balls.

Such mouth-watering thoughts hastened my step through the entrance of the Walled Garden in the village. This had originally been a source of fresh vegetables for Applecross House, but had fallen into disuse at the turn of the century. The garden and attached building was bought in 2001 by locals John and Elaine, who laboured mightily over the next few years, to recreate the splendour of the Victorian original. They cleared impenetrable brambles and bracken, tilled the soil by hand and gradually reclaimed the genteel walled enclosure from the thorny and fibrous interlopers that had run amok in the intervening century of neglect.

In the last few years, the couple who had taken on the

mantle of looking after the Walled Garden were Peter and Jackie, and it was they who all and sundry had recommended me to chat to about creating the perfect crofter's garden. As I walked past immaculately manicured lawns, and flowerbeds that looked like a snapshot of a fireworks display, I could see them in the middle distance, crouched over a barrow engaged in deep discussion about the plants within.

They both straightened as I approached and, with Jackie stepping forward first, introduced themselves. She looked like the absolute clichéd image of a gardener, floppy hat shading a kindly face, with red-apple cheeks, eyes creased at the corners with laughter lines, and a shy smile accompanied by a surprisingly firm handshake. I liked her immediately – as I imagine most people do on meeting Jackie – and rather wanted her to be my mum. She looked like she could churn out vast pies, bake lovely bread whilst simultaneously making jam, all the while pursing her lips to blow a loose tendril of hair away from her eyes and sporting a dab of flour on the tip of her nose.

Peter seemed slightly more stoic, a tall, generously built man with a greying beard and shiny bald pate, tanned a deep brown by the sun. He too was dressed precisely how a gardener should be – baggy jeans with deeply stained green knees, tucked into scarred green wellies, an open-necked checked shirt, and a huge pair of gardening gauntlets, one of which he removed to shake me by the hand.

'Well, now,' he said, studying me with slightly unnerving intensity, 'I understand that you want to start a wee garden down on the croft in Sand Bay.'

I replied that I did – what's more, one that would feed me come the end of the summer.

Like so many gardeners, their lives dictated by the seasons and the slow emergence of their plants, Pete liked to take his time.

'Well, now,' he said again after a suitable pause, punctuated only by the rasp of one thumb along the edge of his beard, 'judging by the look of you, I wouldn't say you know a whole lot about plants.'

This wasn't said unkindly; it was more of a simple statement of fact. I had no idea how I should look to exude an air of horticultural confidence, but it obviously didn't feature in my make-up.

'Oh, Peter, don't be so rude,' said Jackie, scolding him softly. 'Come along, and we'll show you round. There are all sorts of things we can show you that will be the perfect start.'

'Thanks,' I said, only just avoiding saying 'mum' on the end.

And so began my wondrous tour of the Walled Garden with two of the most delightful guides you could possibly hope for. Pete and Jackie had plainly accumulated vast amounts of knowledge when developing the garden, and walked me through the magical kaleidoscopic ranks of the flowerbeds with justifiable pride.

Everywhere there was vibrant, riotous colour. I had always regarded plants as – quite frankly – rather dull, preferring things that plunged and shrieked and darted. Here though, the reds of geum and lychnis fought a battle with the purples of lavender and verbena, whilst the orange of nasturtiums and marigold vied with the yellow nodding heads of the rudbeckia. In the lower beds, the new vegetables sat in serried ranks, recently transferred from the sanctuary of the greenhouse and staring upwards at the horticultural bedlam above. They looked rather nervous, like new children in the big school. I mentioned this to Jackie.

'Oh, I know exactly what you mean,' she said, her eyes twinkling. 'I have little chats with them all the time, and I get very emotional when it's time to harvest them. They're my little brood really.'

Peter told me that the lettuce were not just lettuce – they were speckled trout, iceberg and red oak leaf, wonderfully evocative names, plainly created by someone with a great deal of time on their hands (i.e. a gardener). I had visions of an elderly gentleman puffing contentedly on a pipe, peering thoughtfully skywards in a haze of sweet shag, his muddied notebook open before him. Perhaps there was an unlikely event involving an oak, a large trout and an iceberg that took place outside his front gate; however, I like to think it was a more meditative process, and by the end of the day he had scrawled the names and was sipping Earl Grey whilst tucking into a slice of Battenberg.

The artichokes spoke of more exotic origins – globe and Jerusalem – whilst the myrtle bush was from Chile, looking swarthy, robust and thoroughly South American at the garden's entrance. Over it all stood the apple trees, branches twisted with age, rheumatic and knobbly and bowed with the weight of annual expectation. The massive leaves of gunnera lined one wall, looking like something out of a science-fiction B-movie, ready to uproot themselves and go on the rampage through the village. Underlying it all were the smells, a heady concoction of perfumes through which flying insects hummed and weaved in a narcotic haze.

By the end of the walk, I was a complete convert. Over a cappuccino in the Potting Shed – the small café nestling at the back of the garden – I asked Peter and Jackie if they could help me set up my own garden. They agreed immediately, and I left them later as they plotted furiously, two artists presented with a blank canvas.

*

Keith and Rachael Jackson's farm was tucked away in the lush north-western arm of Skye, and was reached by a twisting single-track road that wove through low hills, a contour line made of tarmac. Swerving between low walls and over ancient bridges, all I could see around me were green rolling fields dotted with sheep – this was farming country, lacking the fripperies of the coastal road that pandered to the tourists who flocked to Skye in the summer months.

Rounding yet another tight corner, suddenly I was in the drive of the farm with the house right in front of me – in fact, I was heading towards the front door at forty miles an hour. I wasn't particularly keen on crashing the Land Rover through the porch on my first visit, and stamped on the brakes with both feet, crunching to a stop in a cloud of dust. This duly cleared to reveal Keith and Rachael emerging to say hello.

I soon discovered that their unflappable reaction to my entrance was pretty much how they dealt with everything in life. Running a successful croft requires a certain degree of stoicism, a quality Keith and Rachael had acquired in spades over the preceding few years.

Keith introduced himself first, a warm smile spreading across ruddy cheeks. He was a compact, dark-haired, powerful figure, emanating calm competence. Rachael, an attractive blonde with intelligent blue eyes then stepped forward to say hello – with a distinct Australian twang. Seeing my reaction, she laughed and told me of her somewhat tortuous route to a croft on Skye.

'My family were actually from the Isle of Skye many generations ago – I'm descended from the Macleod clan. Although I'm a second-generation Aussie and grew up in Sydney, I travelled here when I was 20, and I've never left. As soon as I got here it felt like I had come home, so I stayed.'

'Aye,' chipped in Keith, 'to think she could have been sipping a cappuccino in a flat overlooking the Opera House, with a rich banker for a husband. Instead she got me.' At this he chuckled, gave a theatrical stamp of his wellies, and turned to lead me into the house.

Over lunch, Rachael told me of a house on Skye where she had stayed a night as a backpack-toting travelling Aussie lass. She had remarked as she left about how she had felt an instant sense of comfort and belonging, a benign ease and familiarity about the place that she couldn't quite fathom. It was only several years later that she discovered that it was the house where her grandfather had been born.

I explained to Keith and Rachael my plan to raise livestock on the croft. Both listened with absolute attention, with Keith asking the occasional short question, and then lapsing back into thoughtful silence. When I had finished outlining my plans, Keith leaned forward, filled my coffee cup and began to talk. It was, very plainly, my turn to shut up and listen.

'Monty, what you need are animals that can cope with this environment almost regardless of what you do to them or where you put them. My guess is that you're not overly versed in raising livestock,' he said with a distinct twinkle in the eye, 'and you need native breeds that – if you simply put them out on the hill – would instinctively know what to do. You've come to the right place, because we've been farming native species here for some time. The best thing to do is take you on a whistle-stop tour of the farm and introduce you to some of the potential stock you should be taking on.'

Minutes later, we were striding through a busy yard, strewn with old machinery held together with twine and tape, with bits of one thing bolted onto bits of another. Recycling is far from a new concept in crofting – every day is a new test of initiative, of how to extend the life of a piece of equipment or jury-rig a new system for feeding, corralling or caring for stock. Years of being forced to use his initiative had turned Keith into the walking equivalent of a multi-tool.

I was somewhat taken aback to see a peacock in the yard, looking incongruous with that gloriously sleek ball gown of a tail, and an emerald neck topped with a tiara of delicate black feathers. He was picking his way through the tyres and rusting machinery, looking rather like a visiting dignitary performing an onerous public duty.

'Keith, why on earth have you got a peacock?' I asked.

'Oh that . . . We swapped it for the wallaby,' he replied, as though it made perfect sense. 'Come on, keep up.'

Our first stop was the pig barn. This was where the indoor pigs were kept. Distinctly cosseted by pig standards, they had warmth, cover, bedding and feed. These were here to be fattened up for bacon – corpulent slickers grunting contentedly, all the while growing streaky under their coats.

'These are a cross between Tamworths and wild boar, and are a very, very rugged pig indeed. These are all from one sow, and you can actually see their genetic make-up in their coloration.'

Sure enough, there was the complete range of stripes, spots, browns and reds that neatly encompassed the mixed heritage of the mother.

'I thought those two might be good for you,' said Keith, pointing to a russet-coloured sow and medium-sized boar, the latter with a distinct ruff of coarse hair in a spiky quiff between his shoulder blades. 'That sow is absolutely beautiful. I'd say for you, as the weeks roll on, she could bridge the livestock-girlfriend divide quite neatly.' He gave a great shout of laughter and turned on his heel saying over his shoulder 'Come and meet the yobs.'

A short walk took us to a patch of woodland nearby, with dark conifers rising from ground that appeared to have been shelled – a dark mass savaged by tectonic forces.

'This is where we keep the outside pigs. These guys live here throughout the year, and are lean and mean. You'll notice when we see them that they have a much wilder look – all shaggy coats and powerful physiques. These are like little wrecking balls, and they're doing what they are programmed to do at the most basic level – find food, have a punch-up, and either hump or bully anything they come across.'

It sounded like a stag night in Cardiff, a view reinforced when the pigs hove into view. Perfectly camouflaged within the dappled wood floor, they appeared en masse, barging and swaggering. We were standing deep in the woods by now, and they swirled and jostled around our legs, causing me to dance and skip like a 6-year-old ballerina performing in front of her parents.

Keith grinned hugely, 'Show them who's boss, Monty. They'll pick up straight away if you're nervous. You need to be top pig.' With this, he eased one of the pigs aside with a firm toe end, and headed towards the gate. I followed in his wake, glancing apprehensively over my shoulder at the rabble behind me.

Next on the list were the sheep – these were no ordinary vacant four-legged balls of cotton wool though, like those we are accustomed to seeing down south. These were Soays, a breed apart. The word Soay speaks volumes about the origins of these tough little sheep. It is derived from a Norse word meaning 'Island of the Sheep', and is still the name of the barren lump of rock – only just over a hundred acres in size – that is thought to be their original home. There is speculation concerning their origins, whether the Vikings brought them over in the ninth and tenth centuries, or whether they found them there on arrival. Regardless of the uncertainty about their lineage, one thing is certain – this is a very old breed indeed. They are unsullied by inbreeding and, rather unusually for sheep, are as hard as nails.

'I raise two types of sheep on this farm,' Keith told me. 'Soays and a more classic commercial breed. The commercial breed suffer about a sixty per cent mortality rate in their lambs due to predation from eagles. The Soays never, ever lose a lamb – they hide it when it is born, and then it moves around between their legs when it is a bit older. These guys have been fighting the war against the eagles since time immemorial – we're talking about similar sets of skills to the antelopes on the plains of Africa here. The commercial breeds just don't have that instinctive response to predators, and the result is a massacre come lambing time.'

That evening was spent in the company of Keith and Rachael and their two children, Tristan and Jessica – both of them happy, healthy and bright as buttons. Long after the kids had been packed off to bed, the whisky and conversation flowed in equal measure, with Keith and Rachael delighting in telling tales of their own misadventures – being chased around farmyards by irate boars, Soay sheep escape bids, and encounters with deranged tourists. The sun had long set, the stars glimmered in a cloudless sky, and the porch light glowed as a tiny glimmer in the dark velvet of the surrounding fields, a well head for shouts of laughter at the end of yet another tall tale as the night rolled on and the sheep grazed quietly in the darkness.

*

As well as raising pigs and sheep, I had decided that I would also have chickens, rather liking the idea of the cosy domestic scene of scratching feet, contented clucking, and regular production of eggs for breakfast. I had to collect the chickens from the encouragingly named Donald 'the Hen' MacDonald, who ran a croft in the tiny village of Strauan, a short drive from Keith and Rachael's farm.

I made my choice from the babbling, scraping hordes before me. It wasn't the most scientific process in the world: 'That one looks nice . . . ' As I did, Donald enlightened me about the wonderful world of the humble hen.

'You know, Monty, there are more hens than people on Earth,' he told me, 'and man has been domesticating them for 8,000 years.' Leaning on the gate, he gestured at the multicoloured flock with considerable pride. 'There are 150 different species, they'll lay about 250 eggs a year, and can live for up to 20 years.'

He turned to me with a twinkle in his eye. 'And one more thing, laddie. The hen is the closest living relative of the tyrannosaurus rex, so mind you don't turn your back on one.'

With a chuckle he turned away from the pen, returning moments later with a pile of cardboard boxes. We packed the hens into boxes and hefted the coop onto the trailer I had borrowed from Mike. The long drive home involved crossing the mountain pass, and we all enjoyed a thrilling journey back down into the village, the trailer swerving and rocking behind me, with the hens clucking in alarm as they experienced G-force at altitude – something I imagine is an entirely new experience in the world of hens generally.

Back at the croft, Sam, Andy and I moved the coop over various walls and through tiny gates, sweating and swearing, each slip greeted by the noise of tumbling feathery bodies and a cacophony of clucks. Finally settling the coop on the ground in the pen, I eased the hatch open, and to my immense satisfaction the first hen immediately poked her head out and began to scratch at the ground. I felt inordinately proud of them, and stood with a proprietorial smile on my face. More hens appeared, and a feeling of vast calm settled upon me.

I really should have learned by this stage that when I'm in this mood Reuben appears to sense it at some deep level, and sees it as his bounden duty to snap me out of it. My first indication that things were about to become quite exciting was the brush of a hairy back against my thigh, and a large black head gliding forward sporting a delighted grin that exposed white incisors – an alarming sight for even high-altitude adventurous hens like these. He moved over the ground with demented acceleration, eyes wide with delight.

Reuben had of course never seen a hen in his life before, but there is a fairly basic response within dog DNA when faced with eleven clucking, squawking, flapping, fat objects. The hens spotted Reuben, and ran furiously in small circles occasionally hopping pointlessly off the ground. Reuben was now in what combat pilots would describe as a target-rich environment, and leapt in with relish, creating a whirlwind of dog, spiralling hens, feathers, dust and bracken.

Several traumatic minutes later, Reuben was locked in the Land Rover with an assortment of tail feathers clutched in his snout, the traces of a dreamy smile still curling his lips. The hens were now in a variety of locations, with some on the bothy roof, some jammed into tiny crevices in the wall staring glassily into space, and some sprinting over the dunes towards the horizon, like plump matrons hitching up their skirts and racing towards the last bun at the fête.

It took an hour to round them all up and restore a semblance of order. Reuben had by then been released from the Land Rover and lay dozing in the sun planning round two, with his back legs twitching to the accompaniment of contented snuffles. The hens were once again scratching at the earth, with only the occasional nervous glance towards the gate. I was sitting in front of the bothy, nursing an uncharacteristically early whisky with trembling hand, the enormity of the task ahead of me suddenly sinking in.

*

The following morning I stood expectantly on the tarmac road leading to the bothy track, squinting into the middle distance. I saw the glint of an approaching vehicle, and moments later heard the roar of a labouring engine. Keith and Rachael's battered estate duly appeared over the rise, towing a large galvanised trailer. They waved enthusiastically through the windscreen, before turning off onto the track and coming to a halt. The car doors flew open, the entire family emerged from the dusty interior, and Keith strode towards me, hand outstretched, casting his eyes skywards as he described the journey.

'Aye, Monty, it was hell – the car started to make some very strange noises coming over the pass. Bear in mind I had the family, plus camping gear, plus two pigs and eleven sheep in the trailer – a breakdown through the hairpins would have been no fun, I can tell you.'

As if to back him up, the trailer emitted a clunk followed by some very annoyed grunting and shuffling noises as it rocked on its axles.

'Best we get these chaps out onto the croft as soon as possible. I'd hate for the first thing two hot and bothered pigs see when they thunder out of the trailer is their new owner.'

We wheeled the trailer carefully down the track by hand until we reached the sheep enclosure, lifted the door and released the Soays. Leading the way was the ram, only two years old but cocksure and defiant, with horns sweeping in elegant curves around his skull. He was swiftly moved along by the four ewes that followed him down the ramp, calling for their lambs to follow, all big brown eyes, tiny delicate heads, and slightly wobbly legs. Age-old instincts immediately kicked in, and the ewes ushered the lambs straight towards the bracken where they immediately disappeared, their dun coats merging with the shadows and fronds. As they vanished from view, the flock instantly stopped calling, and within seconds it was as though they had never existed. As a raw example of honed survival skills, it was very impressive. This was with the exception of one tiny dark lamb, who was still coursing back and forth along the edge of the bracken, bleating pitifully. The undergrowth mysteriously parted, the ewe's face appeared, and the lamb plunged into the gap as though diving into a green sea.

Keith was already manoeuvring the trailer towards the pig enclosure, stopping a little way from the gate to once again lower the back door. The pigs slid down the ramp, sniffing the air and peering about them cautiously. Keith moved ahead, calling gently to them and wheedling them into the enclosure. Keith shut the gate behind them, and we both leaned on its creaking timbers, consisting as they did of the hammered together remains of the old roofing frame from the bothy.

The first thing that struck me about the pigs was their size – these were chunky beasts, with muscular barrel-like bodies and surprisingly delicate limbs, like an overstuffed punch bag with a miniature piano leg in each corner. They were covered by coarse, russet-coloured hair, and their necks were immense, rising in a slab of muscle towards two ears that swivelled and twisted like radar, scooping up noise from their surroundings. Their noses seemed to have a life of their own – pink, twitching and immediately hoovering the ground of the enclosure, occasionally being driven into the dirt with a muffled grunt, a heave of the neck and an embedding of pointed trotters. They were perfect digging machines, and within minutes they had turned over a small patch of ground and were snuffling around for bulbs and roots.

I named one of them Gemma, after my niece. The other I called Doris – because, for some reason best known to myself, I had always wanted a pig called Doris. They were, quite simply, beautiful.

It was only the next morning that I began to suspect that Keith had given me a psychotic pig.

Gemma was a delight, jogging towards me on stumpy legs, uttering grunts of pleasure at my mere presence. Having stopped before me, she would gaze upwards in complete adoration, hopping up and down on her two front trotters as if the excitement of seeing me was all a bit too much.

Doris, on the other hand, was obviously a pig with issues. She would circle in the background, grunting with disdain, occasionally glancing in my direction with a curl of the lip and a toss of the head. On the first morning that I entered the enclosure to feed her, I made the mistake of wearing flip-flops and shorts. You really have no idea of the acceleration you can achieve in flip-flops until you are charged by a baying pig, and I surprised myself with a sprint that showed great pace, splendid high knee lift, and culminated in a hurdle of a low wall that would have done a sprightly teenager credit.

Plainly this could not go on – particularly as Keith had urged me to establish a pig hierarchy at an early stage. I entered the pen the next day in stout wellies and clutching a handy piece of old roofing timber in a sweaty palm. Doris saw me, the red mist immediately descended and she charged with an impressive spray of turf and small stones, emitting an ear-splitting squeal.

As well as having tremendous acceleration from a standing start, Doris had the ability to apply the brakes with equal alacrity. She did so when I came properly into focus – pigs are notoriously short-sighted – and noticed I was in the position of a batsman about to receive a yorker: roof timber waggling in small circles, legs akimbo with feet firmly planted. We stood glaring at each other for a moment, Doris sitting on her backside with her rear trotters sticking comically through her front legs, and me wide-eyed and wound like a spring. Then she righted herself and trotted off without a backwards glance. It seemed that some sort of hierarchy had been established in that single moment and from then on Doris became, if not completely cooperative, at least grumpily respectful.

From this point on, my mornings developed into a set pattern, dictated by the needs of the hens, the sheep, the pigs and, of course, Reuben. I discovered at an early stage that I was way down the food chain, and would clump around furiously in wellies, keenly aware that each task brought the first coffee of the day ever closer.

The day would begin with my face being gently nuzzled, an ear being licked, and a wet nose being pressed against my cheek. Under the right set of circumstances this would have been splendid. A swift peek through a half-open eye, however, would reveal the dog's delighted face at close range. Swinging my legs out of the bed, I would climb wearily into

71

clothes of questionable cleanliness and stumble down to the seashore, closely followed by the dog. Here, with a yawn and stretch, we would both have a morning pee, with me making polite small talk with Reuben as to the nature of the weather, and general plans for the day ahead.

Hearing the door of the bothy creak open was the cue for the rest of the menagerie to demand attention in the most strident terms. The pigs would begin a series of basso snorts, which in turn would trigger off the hens, which would then bring the sheep running to the edge of the fence. I would therefore walk back up to the bothy under the gaze of a host of hungry eyes, pausing only to put the kettle on before commencing with the morning feed.

This would begin with releasing the chickens. The rooster – which I had christened Dave the Rave due to his large harem – would by now be crowing furiously, and as I approached the coop I would see his face framed perfectly in the metal grille that doubled as a small window. He would glower at me as I approached, the cockerel equivalent of tapping a wristwatch. Throwing open the hatch, I would stand back and watch the hens emerge, invariably led by Dave, who would fluff up his feathers and give a quick experimental flap of his gorgeously glossy wings, before turning to inspect his ladies as they followed him into the morning light.

The hens released, I would then collect a bucket of pig feed and head determinedly for the pig pen. By now Gemma and Doris would have their heads rammed into the bars of the gate, both leaning into the creaking wood like plump ladies pressed against the window of a cake shop. My appearance would elicit shrill grunts of impatience, and I would have to throw a few handfuls of feed into the enclosure to release the several thousand pounds per square inch of pork that was being pressed against the gate. Sneaking in when their backs were turned, I would spread the rest of the feed on the floor, dancing back as they turned and sprinted towards me, rocking and rolling like mini trucks in a demolition derby.

Having survived the pigs, I would turn my attention to the sheep, who would now be bleating pitifully and following me en masse along the fence. Trudging over the field, I would spread their feed and fill the trough, before returning to the bothy to turn the whistling kettle off. Reuben would now be looking at me with gigantic brown eyes, a string of saliva from his jaw glinting prettily in the morning light. Wearily spooning out dog food, I would place it on the floor for him to leap forward, burying his face in it and inhaling furiously, the canine equivalent of bobbing for apples.

One last quick tour of the pens would see all the water troughs filled and the crops watered, before arrival back at the extension.

I would then make a coffee and sit, reflecting on the gentle, undemanding life of the crofter.

30 May

My First Game of Shinty

Saturday dawned as yet another gorgeous day, with mackerel clouds scudding over the hilltops in the distance, a breath of wind bringing a whiff of heather, and the gentle crump of small waves breaking on the beach. I was particularly pleased that the weather was fine, as today was to be my first experience of the ancient and noble game of shinty.

I was first made aware of the shinty team when Bob, the team's coach and the local expert who helped me with my deer fence, invited me to play. When I mentioned the meeting to Sam over a pint, and told him I was keen to play a game, he laughed so hard that beer came out of his nose.

'You do know what shinty is all about, Monty? It's essentially a battle, but with a pointless chap running round blowing a whistle every now and then.'

'Well, Sam, I've played a fair amount of rugby, and it can't be worse than that.'

Sam simply smiled, and returned to his pint. Later I saw him chatting to some of his friends in the corner of the bar, with what I thought was a significant glance in my direction followed by a shout of laughter.

I duly headed over the pass on Saturday morning with a view to watching my first game, deciding that the best plan was to make up my own mind about taking Bob up on his invitation. I would be taking into account the fact that I was now a forty-something, slightly balding man with a dicky back and aching joints. The wasp-waisted young Marines officer was a distant memory, and years of foolishness on rugby pitches combined with a love of diving had turned me into something of a walking physiotherapy lecture. Every morning my knees reminded me forcibly that their combined age was 82, creaking and squealing like rusty gate hinges as I tottered to the bathroom.

The village of Lochcarron soon came into view, a pretty sweep of white cottages along a rocky shore with an immaculate shinty pitch at one end. As I parked the Land Rover, I could see that preparation for the game was already under way, with groups of worryingly fit-looking young men limbering up on the neatly clipped grass. Every now and then one of them would trot over to a small white ball and give it a mighty thwack with a viciously hooked club, to send it whistling towards a lunatic in the goal mouth, who would stop it nonchalantly with an outstretched hand. There were two options here. Option one was that the ball was made of marshmallow. Option two was that these people were insane.

Shinty was developed by the Celts in Pagan times as a preparation for battle, and is a game revered for its uncompromising approach. A recent historical study observed that Cú Chulainn – a notable Irish hero – proved his mettle by driving a ball through the foaming mouth of a mad dog, 'forcing the brute's entrails through the other end'. As such, the women of Ulster sallied forth to meet him, many of them in the buff. He was said to be so overwhelmed by this sight that he had to be 'placed in three vats of water to quench the ardour of his wrath'. This is not a story I could ever imagine being associated with, say, croquet. I settled onto a bench by the side of the pitch with a keen sense of anticipation.

By now both teams were on the pitch, and the referee – immaculate in an alarming tangerine-orange outfit – blew the whistle for the team captains to step forward. The crowd numbered around fifty, although they made up for their lack of numbers by their volume. I couldn't make out what they were shouting, but the tone was that of a beery Elizabethan crowd at a flogging, with a general air of imminent bloodshed and possible mortal injury hanging cheerily in the air.

Having chatted to the captains, the ref waved them back to their positions and everyone – the crowd included – hushed expectantly and stared at the small white ball in the centre circle. Two players from each team stepped up, the referee picked the ball and threw it up in the air, at which point all hell broke loose.

As far as I could figure out from the early exchanges, shinty appeared to be a game combining hockey, rugby, golf, illegal basement dog-fighting and bare-knuckle boxing. The shinty stick is sharply angled, meaning that a good clean connection creates a rapid upward trajectory for the ball.

Ideally this trajectory will combine neatly with the onrushing soft parts of your opponent, creating an instant crumpled figure clutching said parts and whimpering pitifully. This gives the striker of the ball a good chance to run ahead and continue the move. Using the stick above head height is not only allowed, it is actively encouraged, the only problem being that above head height is just that – generally it is more effective to lower the whistling, sharp-edged stick to eyebrow level, leading to some satisfying connections during the course of a tackle.

Tackling is achieved by two onrushing players combining at speed into a mass of elbows, sticks and knees, with general thrashing and heaving all round. During one particularly robust coming together, I saw one player fall to the floor heavily, rolling onto his front, presumably to protect various fleshy bits from the human threshing machine overhead. I was somewhat alarmed to see his opponent lean over and assume a position that would be ideal to insert his stick up his fallen enemy's back passage. By this stage of watching the game, I wouldn't have been too surprised if such a move had led to a rousing cheer, and possibly the awarding of a special bonus point. To my considerable relief, the tackler merely hooked out the ball from under his opponents body, and charged off upfield.

And so the game continued, with numerous point-blank shots at the goalie, whose only means of protection seemed to be a slightly stiffer pair of shorts than his team-mates. The final whistle blew, the teams shook hands warmly, and the crowd dissipated, seemingly satisfied that it had been another good game. It appeared that the Lochcarron side had won, although I didn't see most of the goals as I was watching through the gaps in my fingers at the time.

I chatted to one of the home team as he strolled back to the club house. A tall, rangy player, who smiled and introduced himself as Angus, he had struck the ball with extraordinary ferocity and accuracy during the game. He seemed as good a man as any to ask for a quick summary of shinty, and whether having a go would be a good idea or tantamount to suicide. My slight concern was that a lanky amateur running down the wing shouting, 'To Monty – I'm in the clear' in the crispest of English accents would be akin to tethering a goat in the midst of the African bush and expecting it to last the night.

Angus assured me that my fears were groundless. 'Ach, it looks a lot worse than it is. As long as you get stuck in, you should be fine. If you come and have a game for the seconds, we'll make sure we take care of you.'

Angus told me that one of the worst injuries is when the stick is planted in the ground, just at the moment an opponent's stick strikes it, causing it to slide up the handle at an alarming rate and rattle into the white knuckles of the man holding the planted stick.

'Aye, you'll never meet a shinty player who hasn't broken a few fingers,' he said with a smile.

I asked Angus about my main concern – that of receiving a ball at short range in the most delicate portions of my anatomy.

He nodded thoughtfully, and looked rather proud for a moment.

'I've actually scored with my testicles, you know,' he said, staring wistfully into the middle distance. 'It was from a corner and I was briefly unsighted. I remember the most tremendous pain in the nether regions, then everyone applauding as the ball ricocheted into the top corner of the net. It's a goal that people still talk about every now and then.'

Wishing me a friendly farewell, Angus headed off to the clubhouse, leaving me to consider the merits of playing a game where directing an iron-hard ball into the opposition net with your reproductive equipment is considered good form. Before I could make good my escape, I spotted Bob by the clubhouse, who simultaneously glanced at me, smiled and waved me over. I glanced longingly at the Land Rover, before turning to walk towards him.

'Well, young man, what do you think?'

'It looked . . . interesting, Bob.'

Bob smiled. 'And will ye come and play a game for us, then?'

The logical portion of my brain howled at me to say no, to laugh and shake my head, to spend my next Saturday afternoon sipping a latte in the nearby café watching the carnage from over the top of the Saturday newspapers. Sadly, we rarely listen to the logical portions of our brain, and with a sense of disbelief I heard the following words – unbidden – float towards Bob:

'Of course. It'd be an absolute pleasure.'

*

I returned the next week, buttocks clenched in anticipation. To my considerable relief, Bob – plainly a decent and compassionate man – had arranged that my one match would be a practice game, allowing me to charge about ineffectually in a flailing mass of swishing air shots. He had also ensured that the teams consisted of the old, the infirm, the pubescent and several carefully selected players who would not attempt to kill me on sight. I emerged unscathed, and actually enjoyed the experience tremendously. The finale was a series of penalties against a keeper who – in his own modest summary – was the best in Scotland. He saved every shot I cracked in his direction, doing so with a nonchalant flick of the wrist before rolling the ball back to me for another go. I felt very much like a 10-year-old playing football with his dad, when he's trying to make every save look harder than it actually is. I drove back over the Bealach having had a splendid afternoon, but resolving that in the future I would stick to rugby, where at least there is the scant comfort that none of the participants are armed.

4

The Storm

Returning to the croft late one evening, I noticed for the first time in my stay in Applecross that the Land Rover was being gently buffeted by winds barrelling up the channel between the mainland and the islands. Glancing out of the window at the sea, still illuminated by soft light even though it was almost midnight, I could see white horses rolling and thrashing on the waves beneath, the ocean no longer the tranquil millpond of the previous weeks. The bracken on the roadside whipped and quivered as the breeze freshened, mini-squalls marking their ruffled passage as they tried to escape the confines of the banks alongside the road to join their brethren racing along the seashore.

Arriving at the bothy, I had to lean a shoulder into the gate, forcing it open as the breeze moaned through the rough wood of the bars. The sheep were huddled in their shelter, almost piled on top of one another in an attempt to share warmth. My only glimpse of them as I rattled past was the glowering face of the ram peering from the shelter entrance, his curved horns a last line of defence as the darkness gathered.

As soon as he jumped from the vehicle, Reuben picked up the feel of the wind, the mood of the impending storm infecting him. As the breeze ruffled his coat, he bounded towards the sea, stimulated by the surf's gathering roar and the seaweed that danced along the edge of the waves, helpless in the grip of the wind. By now the waves were dashing themselves on the beach, withdrawing with a hiss to gather their forces and roll once again up the white sand, as though assaulting the land itself.

Inside the bothy itself there was an eerie hush. The gathering storm, so potent in the wide open spaces of the sound and the beach, was reduced to a frustrated buffeting, with the occasional muffled thump of the door being rattled on its hinges as the wind sought an entrance with whistling fingers. Reuben had by now returned from his foray, and glanced up at me with wide eyes at the rattle of door and moan of timber. His size sometimes made me forget that he was still only a young dog, and there was every chance that this was his first experience of a storm.

The night ahead proved to be a test of the integrity of the bothy and the hastily cobbled-together structures that now sprung from it. Over the next few hours, the extension groaned as though in pain, like a living thing attempting to hold on to the place of its birth. Three times I had to venture out into the storm, twice to check on the pigs as they ran squealing around their enclosure, and once to chase a bucket that rolled and bounced along the beach as though riding the wind to Applecross. Reuben would periodically wake, and growl at the door before slinking back to his bed, somehow wilder than I had ever seen him.

The next morning the storm had blown itself out, with only the foam patches on the beach and the piles of seaweed as a reminder of its full force. It had been mild by west coast standards – 'a bit of a blow last night', as Sam later described it. But it had proved life-affirming to me, sending away the cloying heat and wiping the slate clean from the last few weeks. The build was now completed and, as I walked the beach that morning, picking through the debris, there was a feeling that the exploration of the glorious archipelago around me could at last really begin.

The storm drew some new visitors down from the high ground behind the beach that evening. Perhaps drawn to the shoreline by the dark line of seaweed and jetsam deposited at the limit of the waves or maybe driven down from the hills by the intensity of the wind and rain, a stag appeared from the valley within the dunes, followed moments later by three hinds, stepping delicately onto the sand.

The stag was in his prime, with a magnificent set of antlers splaying into twelve sharp points. This was a magic number in terms of grading stags, with twelve points earning them the moniker of a Royal beast – traditionally only hunted by the Regent. He seemed well aware of his lofty status, surveying the beach before lowering his head to nibble gently on the seaweed, rich in iodine to fuel the battles to come. The sweep of his powerful neck widened at the base, covered in deep curls and dense hair, giving way to a broad chest and muscular legs. The red deer is the largest mammal in Britain, and this dun-coloured monster seemed a throwback to another age, when huge animals stalked the land and man did not hold sway over all around him.

I edged out of the low door of the bothy for a better look, my clumsy movements causing him to turn and stare unblinking in my direction. He was close enough for me to see him sniff the air, moving his head from side to side, ready to call to his harem and send them back into the hills at the slightest hint of danger. Clearly the sight, and more likely the smell, of me was not a cause for alarm, and after a few more cursory sniffs he coughed gently to the herd around him, and lowered his head again to feed. The hinds followed suit.

The waters of the sound also seemed to come to life as May passed into June. For the first time I saw the surface of the channel ruffled in places by shoals of mackerel, a true harbinger of summer. One such shoal edged its way towards the shore, driving a school of bait fish before it, until it was only metres from the back of the surf. I could make out the silver flashes beneath the surface as the mackerel hurled their extraordinary streamlined bodies into the bait shoal, the smaller fish exploding like tiny shards of glittering shrapnel before the speeding missiles of the feeding predators beneath. The commotion swiftly drew in the airborne predators, with gannets and gulls circling the shoal, the latter sitting on the surface as the former dived from the skies. The shrieking clamour of the gulls combined with the thrashing of the water and pattering of the smaller fish exploding through the surface. Within moments it was all over, with nothing remaining but a lone gull still optimistically peering into the water, and my own lingering memory of the fury of the feast.

Inspired by the maelstrom of life on my doorstep, I determined to undertake my first snorkel around the rocks just offshore. The conditions were hardly ideal, with the shallows whipped into a soup of shredded seaweed and suspended particles. However, just as the frenzy of the storm had excited the marine predators, creating a fog of nutrients and small prey through which they prowled on the hunt, so it drew me towards the reef. Dramatic things were happening in the sea off Sand Bay, and I wanted to be in the very heart of them.

Finning vigorously out to the shallow rocks off the beach, I saw the water beneath me begin to clear as the depth increased. Finally the dark rocks of the reef appeared, at first a shadow then as a dense matrix of waving weed and clamouring life.

Space is at an absolute premium in the shallow waters around the British coast and there is a housing crisis of some dimensions on the sea floor, particularly when it comes to a hard base on which to settle. A rocky reef is an ideal place for animals to make their home, a shelter in the storm, somewhere to anchor and wave tentacles in the currents that slip past as a conveyor belt of food and oxygen. The rocks beneath me were covered with life – mussels crowding in on one another, barnacles filling the gaps in between, and starfish lurking menacingly at the margins. The starfish – beloved of any rock pool explorer – are tremendous predators, to the extent that any shellfish that can move will flee before them. The animals that are fixed to the sea floor have had to devise other ways to defend themselves, with even the sloth-like limpet raising itself on a muscular foot to slam the edge of its shell down on any marauding, brightly coloured arm that approaches too closely. What I was looking down on was a battleground, with every animal lurking behind armour akin to some medieval knight, excreting acid to ward off unwanted settlers, or waving venomous tentacles at their neighbours. It is a war for territory that has been fought for millions of years, often on a minute scale but no less ferocious for its lack of dimensions.

Whilst drifting over the reef, I spotted a movement in the sand, a mini-eruption of the sea floor around a distinct oval shape. With an absurd pulse of rising excitement, I realised I was looking at a cuttlefish, the most charismatic and mysterious of all the creatures that lurk in our shallow seas and perhaps my favourite animal in all the oceans.

Cuttlefish come whirring out of deep water in springtime, making their way to the warmer waters of the shallows to breed. Diving down for a closer look, I could see this was a large male, rising from the white sand to hover like a spaceship and peer at me with bright eyes, tentacles waving before it. The eye of the cuttlefish is one of the most extraordinary in the animal kingdom, with incredible acuity and an ability to see even the polarity of light – something that is way beyond us humble mammals. A combination of speed and guile, with a battery of weapons that include a mildly venomous bite and club tentacles that strike with an eighty per cent rate of accuracy, means that the cuttlefish transcends the normal rules for predators. It is a killing machine, stalking, luring, ambushing and devouring prey throughout the reef.

The cuttlefish before me was watching my movements closely, hanging inches above the seabed, the fin around the plump body waving and pulsing, with the jet beneath twisting and flexing ready to propel it into deeper water should there be the hint of a threat. We watched each other for some time, with the cuttlefish edging ever closer, its curiosity and agitation registering on its skin, waves of colour and shade reflecting every change of mood. Soon it tired of me, and glided off into the murk, possibly off to murder some prawns or enfold a crab. I would soon begin to find the remnants of these mysterious animals on the beach, with their cuttlebones washing up on the beach towards the end of the summer, their cycle of life having come to an end with the sowing of the next generation, which would flick and writhe in eggs that hung in dense bunches in the dark recesses of the bladderwrack.

*

The storm signalled a dramatic change in the weather, with sullen grey skies, stiff breezes and rain showers setting in for the next few weeks. The wind direction altered, changing from lazy, tepid spring southerlies into wicked north-easterlies, bringing with them the memory of icy tundra. Evenings in the bothy became rather grim affairs, with me wearing every scrap of warm clothing, crouched next to the wood-burning stove, dinner balanced on my knee and Reuben tucked in a tight, dark mass under my feet. Before going to bed I would strip hastily down to my underwear before cramming a woolly hat down over my ears and crawling into my sleeping bag, falling almost instantly asleep as the rain drummed a staccato beat on the tin roof. My tiredness was due in part to the steady toil of working the bothy, but also to the insipid drain of the cold and damp, my body using deep reserves to combat the enveloping chill. The irony was that we were now well into the early phase of the summer, with long days, and flowers in bloom along the grassy banks leading to the shore. The temperature was spectacularly at odds with the date, and frequently I would decide that enough was enough, and climb into the Land Rover to roar my way to the warmth of the pub or a steaming cappuccino at the Potting Shed café – my haven in the Walled Garden.

The landscape itself changed as the rain fell day after day, at first green and glistening as parched earth and dried heather shone under the downpours, and then becoming a patchwork of colours as the heather flowered in large purple patches on the hills, running riot like some virus infecting valley and hillside. The deep, reddish purple flowers were actually bell heather, as opposed to the true Scottish heather that would not flower for at least another month. Heather is an important plant for the occupants of the moor, acting as food for grazing animals, in particular the red grouse. It also played a key role in the lives of the crofters, acting as bedding – said to be as soft as any feather mattress – floor mats, baskets, brushes to sweep the floor, dye for wool, and most crucially of all as a base for the legendary heather ale. A crofter could therefore feed his stock, sweep his floors, get a good night's sleep, and get rather drunk all from one plant – it's no surprise that heather has become a talisman for good luck in the Highlands.

The streams changed from crystal-clear rivulets tinkling over rocks into angry white torrents streaming down the hills in high-pressure bursts of foam and spray. This was no longer a passive progress to the sea, bending round rocks and obstacles. Now the streams blasted their way downwards, driving flotsam ahead, with their banks sagging and caving behind. The track up to the road from the bothy became virtually impassable, with deep, peaty mud bedding the wheels of the Land Rover down to the door sills, and Sam's bridge the only solid patch in a putrid mire. I purchased some heavily tracked off-road wheels in Inverness, and would make my way to the tarmac in a miasma of spraying mud and noxious marsh gas. Finally I had to venture out and reinforce the bridge, hunched against the wind and rain, skinning knuckles and barking shins as I waddled around the huge rocks Sam had put in place weeks before.

Although for me the weather meant hunkering down and making do, hands thrust deep into pockets and chin buried in my muffled chest, for the vegetables the rain showers could not have come at a better moment. Lettuces that had been limp and ragged in the heat became moist and crisp overnight, opening their leaves to the rain as it drummed upon them. The potatoes sprouted like thick shrubs, jostling for position with onions and beetroot, cabbage and chives. Soon the vegetable patch was thick with life, and I would often find myself leaning on a spade and staring into it, shaking my head in quiet wonder.

I had been fortunate with the weather to date, with my first couple of months showing the west coast at its very best, bathed in gentle early summer sunshine with not a breath of wind. Having yearned for rain, I was now getting slightly too much of it, and had to remind myself that life in the croft was always going to be a mixed bag. Day after day, the sea hissed and thrashed in the channel, rain hammered against the windows of the extension – put in place by a man with the sun warming the back of his neck and not a care in the world. Such a cavalier approach to joinery was now coming back to haunt me, as I furiously stuffed towels into leaking seams that now channelled water down the walls to seep slowly across the floor. I would stomp muddily about my work all day, to close the door in the evening to the drumming of the rain on the tin roof, hammering away as though resenting the dryness within. I would wake the next morning to the same sound, and burrow deep under the covers, dreaming of warm baths, central heating, elegant cafés and flat-screen televisions.

Part of me had yearned for this, the real test of my mettle during the whole enterprise. I had felt strangely short-changed as glorious sunshine greeted every morning and the bothy had sprung up magically under the ministrations of a group of charming and attentive locals. A substantial portion of my rationale for coming to the wild west coast was to explore my own make-up, to see if I had what it took when the clouds closed in, the rivers roared and the ground became a swamp around me. With the arrival of these vicious squalls, every day became a small test, every job a measure of how I would have coped in the long term, in the depths of the bone-cracking cold of a west coast winter. I would be long gone by then, back home in Bristol, as I sipped a cup of coffee, hands wrapped around the mug as I peered out of the rain-spattered window, I imagined a time when the ground would be covered in a blanket of snow, the mountains sheathed in shining ice. A small part of me regretted my absence deeply in a time that was surely the true test of any resident of the west coast.

5

RIB

The best way to travel around the islands of the west coast has always been on the water, something not lost on me as I sat in the bothy, the mystical mountains of Skye looming across the sound. To reach the foot of the first of those mountains by road was a three-hour trip along snaking, single-track roads; by sea, I could be there in twenty minutes.

I asked around in the pub about securing a boat, and was told that the best man to speak to was Peter Fowler, who ran a whale-watching operation on the Isle of Skye. Purchasing a rigid inflatable boat – known as an RIB – had become something of a personal mission for me. I had always, *always* wanted to own one, ever since I first saw Jacques Cousteau on television. Red bobble hat set at a rakish angle, Cousteau set out to explore the forbidden and terrifying world of the open oceans. At this time, the perceived wisdom was that a tentacled monster lurked around every reef corner, that any shark was a bloodthirsty killer, and pretty much every time anyone entered the water a giant clam would immediately close on their ankle. This would invariably happen on a rising tide. The only way out of this fix was to attack the hinge of the clam with a huge knife drawn from a belt wrapped around a worryingly tight pair of trunks.

Cousteau educated me about the wonders of the sea, exploring the globe with a stubbled, Gauloise-smoking crew of Gallic heroes. The archetypal image of Cousteau is at the helm of a Zodiac, with chief diver Falco alongside, kitting up prior to some ludicrous dive to absurd depths. I resolved, aged 10, that one day I would own an inflatable, a red bobble hat and – possibly – a very tight pair of swimming trunks.

It seemed like destiny when I clapped eyes on the RIB for the first time as it bobbed in the harbour at Glendale. She seemed pretty enough, with bright orange tubes sitting atop a sound – if rather patched – hull, and a pleasingly large outboard balanced on the stern. Pete was climbing the steps up the quayside, and he introduced himself with a big smile. He was wearing an all-in-one drysuit, flecked with spray and dried salt, sensibly dressed for the sea beyond the harbour wall.

Pete knew a thing or two about boats. As ever, I was completely at his mercy in terms of integrity and honesty, hardly being an expert on RIBs myself. This was becoming a familiar feeling for me, placing my trust entirely in the hands of a local expert. I had yet to be disappointed.

'She's a lovely boat, Monty,' Pete told me. 'A tad old and weary but well maintained, and the engine is an absolute cracker.'

The engine in question was a sleek, black monster that sat on the stern oozing suppressed energy and excitement, like an adrenalin-vending machine.

'Here we go,' he said, tossing me a lifejacket. 'Let's go and see some seals.'

Hopping aboard, he started her up. The engine coughed into life, then burbled away, trembling and muttering like a caged animal. I slipped my feet into the foot straps and, with a spin of the wheel and a bellow of the outboard, we were away.

I'm not sure what it is – the roar of the engine, the hiss of water under a hull, a fresh sea wind in your face – but I defy anyone, regardless of race, creed, sex or background, not to grin like an idiot when they first go fast in a boat. The bulbous orange nose rose then fell as the boat flattened onto the plane, and we leapt forward towards a distant reef that broke the surface in a maelstrom of foam and swell.

Pete sped towards the seal colony at the northern edge of the reef, slowing respectfully when we were within fifty metres or so. The seals lifted their distinctly canine heads as we approached.

'These are common seals,' Pete told me, 'and round about this time of year they're pupping, so we probably don't want to get too close.'

We hung off the reef for a moment, the boat rocking gently in the swells. The common seal is slightly smaller than the grey seal and has a peculiar habit of arching its entire body as it basks on the rocks, looking like a plump, overstuffed sausage balanced on a speckled midriff.

'Fancy a drive?' asked Pete, sliding off the central seat to make space. I carved a number of tight turns, slightly scaring myself a couple of times, and bobbed around off the colony for a little while. The deal was – as it were – sealed.

I was obviously anxious to get my new toy into the water back at Applecross, and duly undertook the drive over the pass the next day, an emotional process interspersed with some exciting reversing into narrow gaps as various campervans were encountered on the mountain road. This was to be the start of my education in the complex world of reversing large trailers – a process that appeared to defy several laws of physics, requiring counterintuitive turns of the steering wheel punctuated with moments staring into space, looking baffled with both hands mimicking Land Rover and trailer, as frustrated motorists fumed silently ahead and behind.

Parking the boat in the car park at the head of the bay, I walked around it slowly, admiring its predatory lines and vivid colours, aware that before me was a magic carpet, a vessel that would open up the world beyond the beach, giving me access to the white beaches and the forbidding islands that dominated my horizon.

*

The next morning I awoke bright and early. I had called Kayak Mike the day before and persuaded him to come out with me for the boat's inaugural voyage. There could not have been a more suitable first mate – a man who knew the bays and reefs around Applecross better than anyone.

Mike met me at the slipway close to the village, dressed in a high-tech drysuit, over the top of which sat a lifejacket stuffed with flares, safety packs and medical gear.

ABOVE *The village of Applecross, with the inn at its centre and the mountains beyond.*

BELOW *A nervous man and a bewildered dog prepare to head over the Bealach na Ba.*

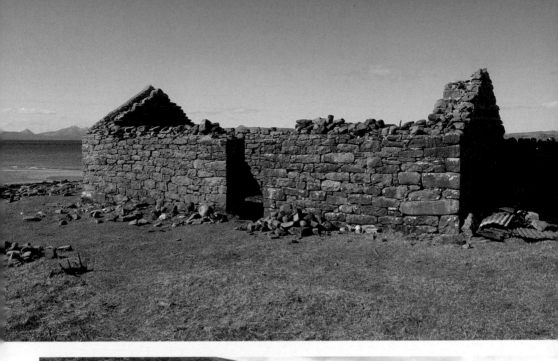

TOP *Rebuilding the bothy commenced under cornflower-blue skies that demanded long lunch breaks.*
ABOVE *Sam (right), Kayak Mike (centre) and I were only too happy to oblige.*

ABOVE *Soon the bothy was a hive of activity, with Sam (RIGHT) always present.* BELOW *When completed, it was basic but cosy.*

ABOVE LEFT The vegetable patch looking lush and verdant.

ABOVE The same patch after the arrival of two delighted pigs.

Under the patient tutelage of Jackie and Peter (LOWER PICTURE, OPPOSITE), I miraculously discovered I had something approaching green fingers.

The arrival of the stock in the bothy was the final piece of the jigsaw. From the triumph of the first egg (OPPOSITE) to ecstatic morning greetings from Gemma the pig, life would never be quite the same again.

TOP A fool, going very fast, immediately before and after he spots a shallow shipwreck.

ABOVE Pied Pipering our way through villages on the Boat Pull.

LEFT Something very nasty about to happen on the shinty pitch.

'Bit of weather coming in,' said Mike with a smile. 'Remember: no such thing as bad weather, only bad clothes.'

I glanced out at the sound, as flat as a millpond dotted with several seagulls sitting gasping in the heat. The weather forecast was good, but the one thing I had learned over the last few weeks was to listen to the locals. This in turn caused me to look down nervously at the shorts and light jacket combo I was wearing.

Mike and I drove to the slipway at Ard Dhubh, a tiny harbour several miles beyond Applecross that had been a traditional anchorage for centuries. There were many reasons why I wanted this to be the first place that I put the boat in the water – deep water, a gently sloping slip and a wide turning circle being some of them. The main one, however, was that there was very little chance that anyone else would be there at this time of day, a key point when reversing a trailer for the first time in a new place.

On arrival at the slipway, I was horrified to see a group of four fishermen standing on the quayside. I recognised one of them as the unusually named Snoddy, with whom I had shared a beer or two at the pub. He had already agreed to lend me five creels – an age-old design consisting of a netted chamber with a hole in the top through which crustaceans would crawl in search of the reeking bait that hung within – ideal for harvesting the rocky shallows near the bothy. Today I would be fishing for bait for those creels, and Snoddy raised a laconic hand in greeting. The entire group turned in my direction, and the reversing began.

Synapses crackled and fired in my brain as hands slick with sweat skidded over a bucking steering wheel. The boat slewed from side to side, the fishermen were now actively watching and thoroughly enjoying the show, and Mike was studying a chart spread on a nearby rock with unusual intensity, a gentle shake of his shoulders the only sign that he was enjoying himself immensely.

In my opinion, you can take every terrifying hazard in the sea, every nightmare from the sounding main, each freak wave and frenzied beast, and all of them pale into insignificance when it comes to reversing a new boat on a trailer down a narrow slip in front of a group of cheery local fishermen. In the end, it all became too much for Snoddy, who stepped forward and talked me through the reversing process. When the stern of the boat finally eased into the water, there was a gentle smattering of applause from the watching fishermen. These being Applecross folk, there was no malice in the group and, as I stepped from the cab, they moved forward to pat me on the back, offer a few tips and assure me that it would get easier with time. I hoped so – from the time it had taken me to get from the top of the slipway to the bottom the tide seemed to have risen at least thirty centimetres.

I thanked Snoddy for his guidance.

'Ach, it's no problem, Monty,' he said. 'It's just that you were in the way of my own car and I wanted to get home in time for my tea.'

We were soon under way, heading out into the main channel at a gentle tick-over, before gunning the engine. The RIB leapt forward, as though spotting the wide open spaces ahead of her for the first time, and we raced towards the island in the middle distance. Glancing towards the open sea beyond the channel mouth, I could see a squall whipping the sea into serried ranks of white horses. I looked over at Mike, who simply smiled knowingly and then looked faintly smug – a vacuum-packed, hermetically sealed, warm-as-toast local staring down the barrel of a summer storm whilst sitting alongside a man dressed for a balmy day in Surrey.

For me, however, the RIB was a considerable consolation, bounding from wave top to wave top, an orange comet blazing a trail with the foaming wake its twisting tail. There was the occasional moment where I got things slightly wrong, the most notable being an exuberant surf down the front of one of the larger waves that had now sprung up, the stern fishtailing behind, with a simultaneous sudden leap in speed as the massive power of the surging water beneath caught us. Sadly I had failed to compensate for the additional acceleration, and was somewhat alarmed to note that the glassy down slope of the wave we were on ended abruptly in the sheer back of the wave ahead. This we hit at exhilarating pace, burying the nose and causing the RIB to come to an abrupt halt. Everything on board kept going, though, including me. This involved a thought-provoking slide down a long plastic central seat, legs akimbo and eyes wide as I studied the rapidly approaching iron-hard centre console. The resultant impact between my nether regions and the console caused my legs to straighten in an amusing manner on either side. The trip continued at a much more sedate pace, the skipper now somewhat ashen and sotto voce. Mike had taken a firm grip moments before hitting the wave, and was now enjoying himself more than ever, plainly having decided that this trip was a combination of an exhilarating day out and tremendous on-board entertainment.

'This is one of my favourite places to kayak, Monty,' said Mike, indicating a wild-looking island, separated from the mainland by a narrow channel. The island itself was a combination of dark cliffs and mysterious bays, topped by grass so green in the strange vivid light of the storm that it looked like emerald icing. 'And over there is where we camp for the evening.'

He pointed towards a tiny bay of pure white sand, with a rocky reef running out at one edge, and a run-down house sitting nestled in the base of the steep slope above the beach itself.

'That's almost a ruin,' said Mike, 'so we actually camp on the flat bit of grass next to it, light a fire and fish off the rocks. Lovely way to spend an evening.'

The bay was an ideal little harbour, tranquil even in the brisk conditions of the squall, and the position of the house was perfect, cupped and protected by the slope behind. Anywhere else in Britain, this house would have been worth an astronomic sum of money, and yet here on the west coast it was quietly falling into ruin.

'I know what you're thinking,' said Mike unexpectedly, 'but for you buying that house would simply be impossible. Sometimes I wonder about our sanity – there's so much potential here and yet we remain mired in ancient politics. Our buildings lapse into ruin, unemployment is rife, youngsters are leaving, and all because we struggle to abandon the old politics of the land. It's almost a second round of clearances – people just can't afford to stay here.'

Despite his pessimism, I couldn't help feeling that the future of this region was in good hands with people such as Mike driving it on with a potent combination of passion and commitment. No ancient laws, no matter how arcane and embedded, can resist the tectonic impact of a new generation of pioneers and local people spawning exciting new enterprises. Surely sanity will one day prevail, and life will be breathed back into shrinking communities.

We arrived at the fishing grounds, a deep channel running between the islands that Mike had suggested would be a good place to fish for mackerel.

Many people have caught great, glittering strings of mackerel on holiday, and we tend to think of them as simple creatures with suicidal tendencies and immeasurable numbers.

In fact, mackerel are part of one of the most successful of all fish groups – the tuna. There are over sixty different species, and all of them have one thing in common – speed. These are the racing cars of the sea, all streamlined flanks and exquisite curves, designed to explode into a silver streak of acceleration and predatory intent upon sighting a prey item. The larger members of the mackerel family are the tuna, trans-ocean nomads that weigh up to half a tonne and are one of the most significant commercial fish species on Earth. They are undeniably beautiful, a dense missile of muscle with tiny, swept-back pectoral fins and huge plates protecting massive red gill filaments like the baffles of a racing car, designed to fuel muscles with oxygen in the heat of the chase. Just before the tail, itself a neat curve with a stiff scimitar-like blade, are a series of smaller protrusions which break up the turbulence of the water passing over the body and make the passage of the streamlined body through the ocean that little bit more efficient. Perhaps the greatest adaptation to speed lies within, a network of blood vessels called – in a rare moment of scientific emotion – the *retia mirabilia* ('miraculous net'). This allows the blood that reaches the key muscle groups to be warm, maintaining the fish in a permanent state of readiness, like a warmed and stretched athlete moments before a race.

I throttled back and immediately the boat settled in the waves, rolling and pitching before gently turning abeam to the wind. This set up a drift down the channel, and Mike and I lowered our strings of brightly coloured feathers into the deep, dark blue water.

I leaned over and turned on the fish-finder, purchased at a considerable price from a chandler's in Inverness. This was the magic box that would lead me to great flickering shoals of mackerel, with which I could fill my creels, fill my stomach, and feed my pigs. (The latter was something of a surprise, but was yet another gem from Keith, who had experimented with various foodstuffs and found that his pigs went into squealing, trotter-tapping ecstasy when he fed them salted mackerel.)

To my considerable delight, the fish-finder immediately started beeping at me excitedly, a noise accompanied by numerous images of electronic fish sculling their way across the screen. It looked less like a rocky seabed in Scotland than an Indonesian coral reef in the midst of a mass spawning event.

By the time I had prepared my rod with trembling hands, the fish-finder was trilling like a hysterical canary. I hurled the string of multicoloured feathers into the water, letting the line run off the reel beneath my thumb, into the midst of what the fish-finder told me was a seething mass of predatory life.

The next hour proved to be one of the more frustrating in my life. My mackerel feathers remained untouched – I might as well have been lowering down the complete chicken for all the bites I was getting. The size of the mackerel shoals beneath was indicated by the size of the image of the fish on the screen, and a steady procession of obese, slab-sided monsters waddled their way from left to right, passing directly through where my feathers bobbed in the water, propelled by my wildly pumping arms overhead. After half an hour or so, I leaned forward grumpily to turn the fish-finder off.

Mike was more stoic, having fished these waters many times before. He knew that the fish were there all right, but that they had other things on their minds than a string of feathers.

The mackerel in the western Atlantic – which were the ones I was trying unsuccessfully to catch – move inshore in the spring months to breed, with the females releasing buoyant eggs in their millions. It is only when the spawning is in its latter stages that feeding truly begins, and at present it was slightly too early for them to be feeding in earnest. It would be a question of waiting – only a few weeks, possibly even days, before really reaping the harvest beneath our corkscrewing hull.

By now the waters were being whipped into vicious little wavelets that were clashing with the muscular swells running through the channel from the open waters of the Atlantic beyond. This was creating very lumpy seas indeed, and when I suddenly found myself fishing uphill – my line heading at a sharp angle into the crackling wall of an oncoming roller – a single glance between Mike and myself said it all. We hauled in the lines, gunned the engine, and headed for home.

*

The next trip out in the RIB proved to be even more exciting, very nearly resulting in a messy death for all on board. For the first time since my arrival in Scotland, two friends had come to visit Sand Bay. Ian was a property developer from Aberdeen, a tall, languid figure with a dry sense of humour and a fanatical interest in diving. Andy, an old friend from the services, had served in the Parachute Regiment and, like so many of his airborne brethren, was short, powerfully built and insanely energetic. He spoke in a thick Scottish brogue, particularly when drunk. During times of severe inebriation, his rich imagination resulted in a stream of ideas and plans for heroic expeditions – generally related with maximum volume and minimum punctuation.

As soon as they arrived, I decided that they should sing for their supper, pressing a shovel and saw into their hands and pointing them to the front of the house where the beginnings of a path was being laid. Ian was somewhat bemused by this development, although Andy was delighted, attacking the ground as if it were a living thing, shirt off and muscles rippling in the morning sun. He was that special white of the true Scotsman, so pale it bordered on blue. His lineage was also betrayed by the fact that, within twenty minutes, his back had gone a vivid lobster red. Presumably within an hour he would have started blistering and then burst into flames; this was avoided by the shirt going back on, to the relief of all concerned.

Having constructed the path in record time, we were soon sitting on the sofa drinking tea. Whenever visitors appeared at the bothy, there was always a natural gravitation to the extension, and I had found that things tended to develop along fairly predictable lines from this point on. Initial exclamations of pleasure at the view would be followed by polite chatter, with the visitor's eye always drawn back to the channel and the mountains beyond. After a civilised period, we would all fall silent, staring out of the window and thinking our own thoughts. In my opinion, this was a splendid way to spend an afternoon, watching your guests settle into their own imaginations, their bodies relaxed and their faces calm as they stared into the middle distance, minds already drifting over the channel and soaring over the peaks of the mountains, the bewitching view launching trains of thought and igniting dreams. The effect of the view never seemed to fail, with everyone settling into a tranquil state as they devoured what lay before them.

Andy snapped himself out of the reverie first, and demanded a trip out in the new boat.

'We certainly haven't come out here to build you a path and stare out of the window – let's go out for a dive.'

I had not had the boat long, and had reservations about diving from it so soon, but they would not be dissuaded. They gulped down their tea and ushered me out of the door, Andy hurling diving kit into the back of the Land Rover and climbing into the passenger seat alongside me.

I dropped Andy and Ian off just outside the mouth of the harbour, both of them rolling into crystal water, the seabed clearly visible beneath them, an intriguing patchwork of white sand and dark kelp. I followed their trail of bubbles as they meandered contentedly along a rock wall, the engine muttering behind me, the surface of the sea glassy, and the sun on my back. All seemed well with the world.

It was only when they surfaced that life became interesting. They emerged slightly earlier than I had expected, mainly because Andy's drysuit had sprung a leak. He had essentially spent the last thirty minutes of the dive swimming round in a large bag of icy water, gradually going numb from the feet up. He had hung on grimly due to the presence of huge numbers of scallops scattered on the sea floor, stuffing them into his pockets and straps to finally emerge, even bluer than normal, holding great fistfuls of their beautiful, channelled shells in his gloved hands.

Having recovered the divers, I began to head for home. Suddenly the madness of my actions struck me. Here we were, three men in a powerful boat on a glassy calm day, about to return to the harbour for a feast of hand-caught scallops. Surely it would be foolish not to take the RIB for a turn around the paddock, to end the day with the wind in our faces, the bellow of the outboard in our ears, and the adrenalin surging through our veins.

Turning to tell Andy and Ian to hang on, I gunned the engines and swung the bow around, towards the open sea. The RIB leapt forward, the engine snarling, before rising into a high-pitched roar to drive us on, the hull hissing beneath us and the horizon ahead. We were young, we were free, we were crazy, we were heading at great speed for a shallow shipwreck.

Just outside the entrance of Ard Dhubh harbour, a wreck breaks the surface of the water at low tide, sagging against a reef, draped in seaweed and craggy with rust, a monument to a distant error of judgement by the skipper. This wreck has been long forgotten and – crucially – is unmarked by buoys or flags. It lies just outside the safe channel, which is clearly indicated by metal poles embedded in the reef. Only a fool would speed along outside the channel so close to shore.

By an extraordinary coincidence there was not just a single fool nearby, but three. Ian had by this stage assumed a position lying on his back in the bow, facing me as I grinned my way through the gears, taking numerous pictures as we scythed through the clear water.

This means that the actual moment of me sighting the wreck was recorded for perpetuity. One image shows the carefree grin of a man intoxicated by speed, the next is that of the same man viewing his onrushing demise in the form of a very large immovable metal object, just below the surface and directly under the rapidly moving nose of the RIB.

Andy spotted the wreck a second after I did, shouting a traditional Scottish oath along the lines of 'Wheyjaohmafook' whilst gripping the lines on the tubes. I span the wheel, cranked back the revs and braced myself. The unfortunate Ian – wedged spreadeagled on his back at the bow – simply glanced up with wide eyes, his expression much the same as a heifer being slowly led into an abattoir.

Despite the drop-off in speed and our swift turn to starboard, we hit the wreck with a mighty clang, the roar of an outraged engine and a sudden deceleration. By now I was squinting through slitted eyes, holding my breath with my jaw clenched. Much to my surprise I found that I wasn't spinning through the air like a fleshy Catherine wheel, but actually remained standing at the console. After a moment's contemplative silence, Andy shifted at his seat on the tube, glancing up at me with the beginnings of a smile.

'Well, that was a fairly close one,' he said. 'Are ye trying to rub me out in my prime?'

Perhaps the sea conspires at times to warn the foolish mariner of his own mortality, perhaps a gentle tap on the shoulder is occasionally required to indicate that we have overstepped the mark. Whatever the cause may be, we had timed our arrival at the wreck with precision. The state of the tide meant that the water was deep enough to allow the hull to pass over the wreck unscathed, but shallow enough to catch the skeg underneath the wildly spinning propeller, bouncing several hundred kilograms of outboard into the air and nearly ripping the transom off the stern. A quick inspection revealed no damage bar a shiny scar in the metal of the skeg, and a few wobbly screws where the transom met the hull. I turned the wheel and burbled slowly back to the harbour, with an unwritten agreement already taking shape in the boat that this incident was probably best left unmentioned when we bought our first beer in the pub that night.

27–29 June

The Boat Pull

In the second week of June, posters started to appear throughout the village seeking volunteers for what was grandly described as The Boat Pull. The accompanying image was of a lifeboat on a trailer, festooned with bunting and sponsors' logos, with a road twisting away over the hills in the background. There was a contact number on the poster outside the village shop, and I scrawled it down on the edge of a newspaper pressed against the noticeboard as I juggled shopping bags and car keys.

I rang the number from the village payphone, and a soft voice answered.

'Hello, Alfie here.'

'Hi Alfie, it's Monty Halls. The guy who has done up the bothy down on Sand.'

'Ah Monty, yes . . . hello. What can I do for you?'

I told Alfie how much I wanted to take part in the boat pull, all the while not having the faintest idea what was involved. He remained quiet throughout the majority of the call, concluding proceedings by saying that he would pop round to explain what the boat pull was all about. If I remained keen after that particular conversation – 'and there are no guarantees of that,' he noted with an ominous chuckle – then I was in.

Alfie turned up that evening, rattling into view in a hard-backed pick-up truck, the workhorse of all the local fishermen. On opening the door to him, the first thing that struck me was his size – although he was only medium height, his shoulders were massive, with thick arms stretching the sleeves of his T-shirt tight, great forearms and powerful hands. Like so many big men, he was surprisingly quiet, and stood politely on entering the bothy before I invited him to take a seat. Over a cup of tea, he explained precisely what I was getting myself into.

'It stems from an idea I had in the pub one night,' he told me. 'I was drinking with some friends, and we decided that we'd drag a boat over the Bealach na Ba to raise money for the RNLI – just because it seemed a perfectly logical thing to do with several pints inside us.'

Having launched a number of expeditions on a similar basis, I understood completely.

'Well, amazingly, we actually went through with it shortly afterwards. We had a horrible time of it, as you can imagine, but raised a fair amount of money for charity. That was that as far as I was concerned, but more beers followed over the course of the next year or so, somehow the pain and misery of the first boat pull was forgotten, and we decided to do it again. This time we thought we'd do fifty miles over three days. That was seven years ago, and we're still doing it once a year. We've raised over a hundred thousand pounds for various charities, and the plan is to lap Scotland completely on the coast roads over the next ten years or so. Then – assuming our livers can stand it – we'll maybe move on to Ireland.'

There is something about an absurd challenge that is irresistible, something of course that has fuelled British exploration for centuries. I immediately signed up.

There was another powerful motivating factor for me joining the team: my great friend Jason Ward. Twenty years before, Jason had become one of my closest friends in the Marines. In the brutal twelve months of our training, he was a regular source of hilarity and inspiration. He was a magnet for misfortune, injuring himself in exciting and original ways on a near-weekly basis. One of his more memorable efforts was hurling a thunderflash – essentially a giant firework – at the approaching enemy on an exercise, only to have it tangle in his sleeve and nearly remove his hand. Wherever slapstick mishaps and Ealing comedy pratfalls were required, Jason would be there. Come the thirty-mile march – the final test to gain the green beret of the commando – Jason was a shambling mass of bloodstained bandages, slings and elastoplasts. His arrival over the final hill was such a moving sight – a staggering, limping, shuffling ruin of a man driven on by indefatigable will alone – that the commanding officer of the training wing burst into spontaneous applause.

As his career progressed, Jason became hugely popular with his men. He was a soldier's soldier, who loved nothing more than the rigours of the outdoors, the merciless cold of the Arctic or the festering heat of the jungle. He was intensely loyal to those under his command, placing their welfare and security above his own, and becoming in the process the epitome of everything a Royal Marines officer should be.

It was just such a mindset, many years after training had been completed and after several promotions, that saw him insist on being on board the leading helicopter at the start of the Iraq war. This was not where he was supposed to be – military doctrine preferring that the senior officer on an operation arrives after the target has been secured. Jason found this outrageous and, in a furious argument with his superiors, had finally shouted, 'I have a moral obligation to lead my men!' before stalking off into the night.

He was therefore delighted when the order came down from on high that he could indeed be in the first helicopter. Having addressed his men, he prepared to board, smiling that broad smile and remarking proudly to a friend as he climbed up the ramp beneath the thrashing blades, 'You know, at this moment there is nowhere on Earth that I would rather be.'

The crash, killing all on board, has been the subject of numerous enquiries, but the findings were of little consequence to me personally. Jason was gone for ever, a Viking burial in a ball of flame whilst leading his Royal Marines into battle.

Jason had been an enthusiastic sailor for the last few years of his short life. He had frequently press-ganged me into becoming his crew, our progress marked by roars of laughter and a certain level of erratic tacking. A West Country boy, the tiny port of Salcombe in Devon was the place where he seemed most at home, with several balmy summer weekends under a billowing sail that will live long in my memory. The RNLI at Salcombe offered to scatter his ashes at sea, doing so one long afternoon in a sheltered cove, the blue sea shimmering in the morning light and gulls wheeling overhead. Their quiet, dignified respect for the remains of my friend meant the world to those of us on the lifeboat who had come along to say goodbye. Such things are not forgotten, and I was delighted to put my shoulder to their fundraising wheel as a small gesture of gratitude.

*

The boat pull that year was due to be based around Thurso in the far north of Scotland. This is a truly wild stretch of coastline, dotted with small fishing villages and battered by huge Atlantic rollers, all of which have had a run-up of several thousand miles before crashing into rugged cliffs and white beaches.

Having cadged a lift to Thurso, I checked into the team hotel the night before the event, feeling suddenly nervous about the challenge ahead. I had visions of the team consisting entirely of swarthy fishermen with oak beams for shoulders, piledriver thighs, hands like hams, and a low tolerance for callow Englishmen who didn't smell even slightly of herring.

I was therefore delighted to note two things on my way to the bar. The first was my initial glimpse of the team through the window: they seemed to consist mainly of men of all shapes and sizes in comfortable middle age. The second thing was the size of the boat, sitting behind Alfie's truck on a small trailer. I had harboured visions of a lifeboat of considerable tonnage, towed up snaking hills by a sweating team of gurning fishermen, accompanied by the skirl of the pipes, all in all a heroic scene of biblical endeavour. I was therefore surprised – and relieved – to see that the boat was more akin to a dinky toy, approximately three metres long and made of plywood, based on the hull of an old rowing boat.

I walked into the bar considerably cheered. My presence seemed to lower the average age of the team by at least a decade, so how bad could it possibly be?

The man who walked out of that bar several hours later was somewhat wiser, if considerably wobblier. I had been introduced to the joys of wearing a kilt (very liberating it is too), made a woeful attempt to play the bagpipes, and had a vague recollection of trying to drink a glass of whisky through my nose. I was anticipating a hangover of mythic proportions.

Chief amongst my tormenters was Davy Seal, a skeletal pig farmer from Applecross who seemed to have an endless tolerance for whisky. His right-hand man was Gavin, a retired policeman from Dundee, as hard as they come but with a razor wit worthy of Noël Coward. He massacred me in conversation, then murdered me with generous measures of single malt.

I was delighted to see the next day that everyone was as hollow-eyed as myself, mumbling darkly as they took their places on the frame around the boat for the initial pull – a mere nineteen miles on this, the first day. Jock – the only team member younger than me – was on the bagpipes, the aim being to lift the mood of the rabble now gathered around the boat at the start. The road stretched ahead, a morale-shattering ribbon of tarmac stretching over the distant horizon on the gently undulating moor.

'Come on, then,' said Alfie, himself looking somewhat ashen. 'Let's get this show on the road.'

With an audible groan, the team leant into the frame around the trailer, and the boat creaked forward. The groan settled into a grim silence, until Jock struck up on the pipes – a heavenly sound when sober but, when hung over, the equivalent of having a screwdriver inserted slowly up your nose and then spiralled enthusiastically into your brain. The team winced their way through the first couple of miles, before lapsing once again into the stupor of shared pain.

I was surprised to see that Davy was wearing a thick woollen jumper, particularly as it was a very hot day indeed. His role for the day was to accost passing motorists for money, presenting them with the alarming sight of a crazed man wearing an ill-fitting kilt, with what appeared to be a pair of snow-white twiglets sticking out beneath instead of legs, waving cars down and then breathing neat alcohol fumes through the window whilst rattling a bucket under the driver's nose. They all paid up, as the alternative was driving off well over the legal limit through the inhalation of Davy's breath. The collection of money meant that he ran twice as far as everyone else, and I got quite used to the sight of him sprinting past, bony knees pumping, in pursuit of some wide-eyed motorist. The fact that he did the entire thing in a jumper that appeared to be made of a hollowed-out yak was all the more remarkable. By the end of the day, I had concluded that Davy was a physiological one-off. Pig farming's gain was British athletics' loss.

Dragging a boat over winding roads on the very northern

tip of Scotland has its benefits. The first is that the pace is inevitably very slow indeed, so there was plenty of time to take in the view. The boat and trailer did, however, have the alarming characteristic of attempting to overtake the team on the steep downhills. Not possessing a brake, this meant that all of the team members would have to take a firm hold and dig their heels in, kilts billowing and feet scuffling. The implications of letting the trailer go did not bear thinking about – with the entertaining prospect of the lifeboat arriving in the next village at seventy miles an hour, possibly accompanied by two or three hungover men screaming hysterically with their kilts around their ears and smoke coming from the soles of their walking boots.

During the slower moments of the journey, I was able to breathe deeply and take in the scenery, some of the most beautiful in Scotland. The steep dark cliffs of the northern coast are extraordinarily rugged and stark, as though they have only just been thrust through the water's surface by volcanic forces beneath. Gannets and gulls nest on their vertical walls, whilst hundreds of metres below are massive explosions of spray as millions of tonnes of water meet sheer rock, the first landfall since the American coast thousands of miles away. In stark contrast to the bleak and forbidding cliffs were the beaches, elegant curves of golden sand with perfect sets of waves arriving to rise into glassy faces, the wind whisking the tops off in clouds of spray, before they would topple in on themselves in a clean break that would be utterly irresistible to any surfer in the world. Indeed the pro surfing circuit now includes Thurso on its route, the town being filled once a year by bewildered straw-haired Californians trying to find an internet connection.

Jock had a neat sense of theatre, and would charge off ahead seeking out a rocky promontory or a dramatic backdrop. He would clamber up, pipes held under one arm, and then strike up as the team passed. Legend has it that one of the reasons pipers always march when playing is that they were the first people targeted in battle as an army's bravery and morale could be lifted by a single song – 500 pipers in the First World War paid the ultimate price for their influence on their regiments in the thick of the fight. There is also the slightly less respectful theory that pipers always walk because they are trying to get away from the bloody awful noise they are making. Personally I found it a glorious sound, with Jock's music soaring across the beaches and echoing up the glens. We leaned our shoulders to the trailer, causing the lifeboat to increase in speed imperceptibly, a group of middle-aged men lengthening their stride and standing a little taller, moved by an ancient call to battle.

During a pause in one of the villages, Gavin pointed out that it traditionally took seven years for a piper to learn his trade. Ancient legends had it that apprentice bagpipers would be taken away by the fairies, and only returned after seven years' captivity when they could earn their release. 'Mind you, Jock was having such a good time with the fairies that he decided to stay an extra year, which is why his piping is pretty good,' said Gavin, just loud enough for Jock to hear.

The arrival of the boat in the small villages en route had a quite miraculous effect. Drawn out of their houses by the music, the residents would run after the trailer, filling buckets with coins and notes, slapping the team on their backs, and wishing us well. 'You must understand what the lifeboats mean to these communities,' Alfie told me after one particularly generous reception. 'The lifeboat brings home their fathers, their sons, their husbands and their brothers when the storms rise up. It's their lifeline and sometimes their only hope when the slates are being blown off the roof and the rain is rattling the window panes. It's amazing how much money we raise on these trips, but every penny is well spent as far as these fishing communities are concerned.'

And so the miles slowly passed under our feet, with entire village schools turning out to cheer us on our way, the children occasionally following us as we Pied Piper'd our way through valley and over hill.

As we approached the final few miles for the day, I noticed an unmistakable increase in temperature from beneath my kilt. My inner thighs had been swishing against each other since we had started six hours before, and appeared to have reached some point of critical heat, resulting in a clear message that all was not well. My gait was becoming increasingly bow legged – no mean feat when pushing a lifeboat, even a tiny one, up a steep hill. Some adolescent tittering from the stern was my indicator that the rest of the team had spotted my discomfort, adding to the shame of a man experiencing the extreme end of the dry chafing scale.

In the finest traditions of the kilt, I had declined to wear anything underneath. I had taken it that everyone else in the team had done the same – surely wearing any type of underwear was a betrayal of everything the kilt stood for. Compromise on the undies front, and you might as well wear a pink thong for the entire event.

By the time we entered the village that marked the end

of the first day, I was walking like John Wayne, my feet now several yards apart, wincing with each step. After we had stopped, the team immediately delved into various rucksacks within the support vehicle to find warm clothing, the chill wind swiftly cooling down tired bodies. I was somewhat miffed to see an array of cycling shorts and silky lycra on display as kilts were removed, and waddled over to Jock to make my feelings known.

'Monty, there's no way we could do this without something underneath the kilts – it's a myth that you don't wear anything. There would have been a lot of bow-legged battle charges by the Highland clans if that had been the case.'

'Aye, that's true,' chirped up Gavin, 'and it's not just between the thighs. You can always spot a fake Scotsman because he's got a callous on the end of his willy where it's been worn down by the constant rubbing against the Tartan.'

With that, I made my way to the support vehicle in the hope there might be some soothing lotion in the medical kit.

The fifty miles took three days, each concluded with aching feet, gargantuan amber measures in front of flickering log fires, and then a hideous awakening the next morning to lean once more into the trailer, the boat inching forward in a cloud of fumes and weary groans. By the end of the boat pull, I had acquired a blister the size of a quail's egg, my left knee was creaking like a rusty door hinge, and I had promised Alfie that I would not only help him organise the Highland Games in Applecross, I would also take part in what he termed 'the heavy events' – whatever they were.

6

Below the Surface

Once I had moved into the bothy, the prospect of watching the passage of large animals along the channel proved irresistible. Having built the extension, I set up the perfect viewing hide with a telescope on a tripod pointing optimistically at the sea, and shelves packed with well-thumbed wildlife guides.

I still remember the first flush of excitement as I sat in front of the bay window, cup of tea in hand, the expanse of the sound before me. Surely it would only be a matter of time before a broad glistening back broke the surface, and I would lean forward to take award-winning photographs, all from a ringside seat at the best wildlife show in the land.

Weeks later, I was still sitting in the sofa, still scanning the channel and wondering where everyone was. I had seen fleeting glimpses of passing pods of dolphins, the flash of an otter as it rolled into deep water, a speeding minke whale, a pilot whale chasing fish shoals, and a great many mackerel. This may seem a lot, but in terms of my note-taking it took up one side of foolscap, and even these notes were somewhat vague:

14 June – 7.31p.m. Saw passing pod of dolphins. Or porpoises. Around five in number. Or possibly less. May have been hunting. Or just swimming.

When out on the channel in the RIB, I occasionally spotted small groups of cetaceans moving past, but they always seemed to be terribly keen to get somewhere and certainly weren't in the mood to investigate the boat. After three months, I was beginning to despair – it was not as if I felt I had a divine right to see whales and dolphins, it was just that I was beginning to feel positively cursed in my quest. The clock was ticking, the summer was at its height, and every evening in the pub there were tales of dolphin encounters from fishermen and visiting yachts. I would sit and look gloomily into my pint, being gently ribbed by Mike and Sam.

'Perhaps they just don't like the look of you, Monty,' said Mike after yet another encounter had passed me by.

'Aye,' said Sam, 'that's the thing about dolphins. Very perceptive animals – they can always spot a wrong'un.'

My dolphin encounter, when it finally took place, came from the most unexpected source. It was the colossal figure of Big Willie who finally took pity on me, taking me to one side in the bar and pouring me a generous dram. He heaved his bulk onto the bar stool alongside mine and waved his glass in my direction.

'Right, Monty boy, I hear you're after some dolphins.' He leaned forward conspiratorially, his scarred face inches from mine, bent nose meandering down from his furrowed brow. 'I've got just the place. Let's have a dram or two, and I'll tell ye all about it.'

Big Willie is something of an institution in Applecross. I had heard his name in passing, but somehow we had missed each other in the first three months I had been in the village. This was quite an achievement on my part, as Big Willie is very difficult to miss indeed. He weighs in at a comfortable 130 kilos, all of it sitting on a massive frame that is more muscle than fat. Having worked the fishing grounds since he was a small boy (although it must be said that Willie was never a particularly small boy), his shoulders and arms were akin to massive beams, hydraulic grapples geared specifically to lifting heavy pots through the dark fathoms of the channel. This he had done for day after day, week after week, year after year, and even on land he moved as though at sea, gently rolling and rocking his way through the village.

His largesse was legendary. The annual village fishing competition was still many weeks away, and yet already I had been warned of the perils of setting sail on Willie's boat. As a rule, you stepped on board at the beginning of the day and were assisted off at the end. Loaded on board were prodigious quantities of beer, whisky, vodka, port, pizza, dressed crab, lobster and – last of all – some fishing tackle. Willie had a hip flask that was commensurate with the size of his hip – a mighty flagon that held three bottles of port, which could occasionally be seen glinting in the sunlight as he raised it to his lips. Locating the boat in the bays and coves of the channel was a simple matter of following the laughter and singing that echoed around the cliffs.

After one particularly lengthy evening in the pub, a group of us made the short trip to his cottage just outside the village. Here the drinking continued unabated, with myself in the company of several fishermen, feeling very much like a minor-league sipper drinking in the premiership. They talked of the rigours of being a modern fisherman, where the increasing price of diesel and the decreasing demand for their catch meant that even returning with holds heavy with shellfish, having caught to the maximum capacity of their brightly coloured traditional vessels, meant losing money.

'It's coming close to the end,' said Willie with genuine sadness. 'I just can't see how any of us can make a living. The harder you work, the more money you lose. It's easier to stay tied up alongside sometimes.'

And then, as is the west coast way, the evening turned to song, banishing the demons and the doubt. We all warbled away, with Willie the master of ceremonies, until at last he sat and, waving for us all to be silent, announced that he felt a song coming on.

Willie settled in his seat, composing himself. Finally, after several moments of hush, he began to sing.

Big Willie's voice was utterly beautiful, a high tenor ringing round the room, rendering us instantly quiet. He sang 'The Fields of Athenry', a haunting lament that owed more to Ireland than the west coast of Scotland, but summed up the mood of the evening perfectly. As I looked at Willie singing, I was overcome by melancholy at the sight of this colossus of a man, his lifestyle being slowly brought down by forces even his great strength couldn't fight, his voice pitching and soaring as he stared into his past. His great scarred hands rested on his knees as he sang for the generations of fishermen before him, an industry devoured by the global economy.

I would get to know Willie reasonably well, with a steady stream of advice passed on whenever we met in the pub or on the jetty. He was particularly concerned when he discovered that I wasn't catching any lobster, only crabs. This seemed to bother him a great deal, and he would question me closely as to the position of my creels, and offer tips as to how I could increase my strike rate. He actually lifted a few of my creels himself, replacing them carefully in positions he thought would be more productive.

'I was going tae put a tin of tuna in one of them for you, and maybe one of them manuals about survival from that Ray Mears, because there certainly weren't any crab in them. I'd say you need all the help you can get, boy.'

In one moment of candour when the whisky had flowed in amber torrents and we were deep in discussion about his Scottish heritage – something of which he was intensely proud – he remarked in wonder, 'You know, Monty boy, you're a good lad. It's almost like you're not English.' High praise indeed from Willie.

It was clear Willie was a man who knew these waters well, so when he told me that night at the pub where there might be dolphins, I listened carefully.

'If you go tae the entrance of Lochcarron, there are two young dolphins there that will come and have a look at you. They've been there a wee while, and are pretty consistent. That's where you'll find your dolphins – mark my words.'

The very next morning I was heading at breakneck speed towards Lochcarron in the RIB, skimming past caves, coves, bays and beaches that basked in the golden light of a glorious dawn. During my time in Scotland, I had the occasional reality check, a place or sensation that seemed to be the embodiment of everything good about my adventures on the west coast. It was going to be a beautiful day, with the sun rising ever higher over the islands in the main channel, causing the coral sand of the beaches to shine with an iridescent light.

Seals raised their heads as I sped past low reefs still slick with the cold waters of the falling tide, peering at me with dark eyes that shone like precious stones. A group of oystercatchers working the shallows of a sweep of beach were startled into flight by my passage, rising in a raucous crowd to whirr away along the shore, calling and shrieking in protest. The orange bow bounced and skipped ahead, leading me towards a rendezvous with two wild dolphins. I was completely alone, flying low between the islands, with the wake the only sign of my passing as it hissed and writhed behind me. The hull of the RIB skittered and slapped on the crisp wavelets, the engine in full song, a bellowing tenor that seemed to be the perfect expression of my sense of exhilaration.

The narrow channel between Craig and Ardaneaskan hove into view, guarded on one side by a solid-looking lighthouse. Even though it was a calm day, the waters heaved and convulsed over the lip of the rock beneath, a sill that guarded the deep waters of the loch from the main channel. When the tide ran millions of tonnes of water had to rise over this lip, compressed into a small space, spitting and thrashing at the indignity, furious at their departure from the tranquil depths beneath. I slowed the RIB to a crawl at the entrance to the loch, shuddering at the prospect of falling into the maelstrom around me. As the waters passed back into the depths, the surface stilled once again, and I could once again open the throttle, the passing fields and forests that crowded to the loch's shores becoming an emerald blur.

I pulled the boat into the slip alongside Strome Castle, with a view to making some last-minute adjustments to my camera gear and generally preparing for what I hoped would be one of the great encounters of my stay on the west coast. As I drew up alongside the slip, a woman emerged from a beautiful white house that dominated the small cove in which the slip was the main feature. Noticing the RIB drawing ever closer, she wandered over and offered to take a line. And so began a fine friendship, one that would lead to wondrous explorations of the world beneath the waves in this series of coves and craggy bays.

Having tied the RIB to the slip, she introduced herself as Sue Scott, and – noticing my camera kit piled haphazardly in the bow – began to chat to me about underwater photography. It became immediately apparent that Sue knew precisely what she was talking about, combining the wisdom of many years of taking photos with an infectious enthusiasm that communicated itself in every word and gesture. She was in her early fifties, with strong features and a lean physique that spoke of many years of diving in the harsh temperate waters of Scotland. We were soon chatting animatedly, and when she invited me into the house for a cup of tea, I accepted immediately.

The house itself was the old ferry hotel, and stood four-square to the sea, with the road rising at its back, and the garden ending in the furrowed reefs of the foreshore. As we walked along the gravel of the path, I noticed a small gate leading to some rough stone steps that plunged into the crystal-clear water of the cove.

'That's where I start my dives,' said Sue. 'The exit point is further along, around by the vegetable patch.'

The dive Sue conducted on a near-daily basis took her on a gentle ramble through the rock gardens of the shallows of the bay, before following a steep wall festooned with life, then a gently rising seabed of seaweed and smaller rocks that ultimately led to the final exit.

'How long have you lived here, Sue?' I asked.

'Oh, about eighteen years. I've been doing the dive a bit longer than that, though.'

Sue had originally dived here as part of a club, and in doing so had fallen head over heels in love with the old house that was slowly crumbling into ruin alongside the slip. When the chance arose to buy it, she seized it, and had spent the last two decades recording the comings and goings on the reef that spread in the shallows of the bay.

We entered the house, to be greeted by the large and agreeably shambolic figure of her husband, Mike. He was a botanist of some repute, and looked precisely as a leading scientist should, with a beard that spread over his chin and cheeks like a rhododendron bush, and twinkling eyes behind square glasses.

'Ah yes, Monty from Sand Bay,' he said, shaking my hand vigorously. 'We've heard all about you. Let me stick the kettle on.'

The kitchen was beautifully cosy, complete with an Aga in the corner and a cat curled in a box at its base. The cupboards were dark timber, and the shelves were crammed with all sorts of jars and tins. These were filled with herbs, spices, pickles, jams and fruit, a rich array of contrasting shapes and colours. Combined with the sight of the jars was the smell of their contents, a mix of the sea, exotic spices and vaguely familiar herbs, of coffee and warm bread. I wanted to live right there, in that kitchen, for ever, curled in front of the Aga alongside the cat, to periodically don my diving gear and flop into the mysterious world of the bay outside the back door.

We finished our tea, with Mike and Sue insisting on giving me a tour of the house. This proved to be a fabulous warren of corridors crammed with books on wildflowers and shrubs, nestling next to dog-eared guides on the seashore and exotic travel books. On top of the book shelf sat mementos from around the world – chalky shells with intricate whorls and spirals, wood carvings, smooth stones and faded prints. The house even had its own bar, a fuggy den with dark wood panelling and a row of optics shining in the morning light. I had to resist asking for a gin and tonic before settling onto one of the wooden stools to talk of diving expeditions and great adventures. From every window the calm waters of the loch stretched away, irresistibly drawing the eye – a painting that changed every day.

I thanked my hosts, leaving them waving from their front porch with promises that I would return accompanied by my diving kit. Sue had created a bewitching picture of the dive on her doorstep, leaving me with an overwhelming urge to explore it myself. The fact that she had promised to accompany me – the constant gardener guiding me through her beloved patch of reef – made it irresistible.

I headed back into the main channel, scanning ahead for the landmarks Willie had described. The location of the dolphins was – he had assured me – only twenty minutes' ride away and, sure enough, as I swept up the loch the unmistakable outline of the rocky headland he had described in detail appeared before me. Beyond was a small bay, a sheltered cove in which two juvenile common dolphins had taken up residence.

The common dolphin is a very successful animal indeed, having populated vast swathes of ocean around the world. They have very sensibly decided that they will live in areas of warm temperate water, with their distribution also spreading to tropical zones. The only regions from which they are absent are the frigid waters of the poles. They are very beautiful, with an exquisite hourglass marking on their flanks, a combination of pastel shades only revealed as the animal leaps clear of the water. They are smaller than bottlenose dolphins, the other main species of dolphin around the British Isles – reaching a maximum length of only 2.5 metres. They congregate in pods of several hundred animals and, being a coastal species, sightings are relatively common – hence the name. Individuals or very small pods are frequently seen in the northern edge of their range, which – happily for me – includes Scotland.

Willie had told me that the dolphins tended to base themselves at the far end of the bay. I slowed the RIB to a crawl, the engine sputtering behind me. Creeping forward, I scanned the calm surface of the water ahead, the rock wall providing a dramatic backdrop, with an echoing soundtrack of shrieking gulls that spiralled and soared in the blue skies above. It was atmospheric in the extreme, my heart beating and my breath catching at the prospect of the encounter.

I did not have long to wait. At first a glimpse of foam, then another, then two shining backs breaking the surface, charging towards me, an explosion of spray marking the passage of two wild animals rushing to meet man. They seemed frantic in their haste, racing each other to be the first to greet the latest interloper to their quiet home.

I held my course, the boat still moving slowly ahead, the wake a series of ripples spreading from the bow to race away either side of the RIB. Within a moment, the dolphins were upon me, the first immediately propelling his muscular, compact body through the surface, droplets glittering in the afternoon sun, illuminating the markings along his flanks. The second dolphin came in at a more acute angle, changing direction in the very instant of arrival at the side of the boat, an aquatic handbrake turn of breathtaking audacity. He too raced towards the bow, lunging clear of the surface before twisting once again into the open channel, an accelerating dart of energy vanishing into deeper water.

I did what most people do in the first moments of any dolphin encounter - I laughed - and then I scanned the water ahead of me. This time they appeared from the opposite side, running parallel to one another in perfect formation, leaping just ahead of the wake then running directly alongside the RIB, the nearest turning on one side to study me as he sculled past before carving an elegant turn away.

There are a number of explanations why dolphins ride the bow wave of boats. The pure physics are easy enough to explain - a pressure wave being driven ahead of any vessel that offers a free ride. Bow-riding behaviour is thought to derive from the dolphins' distant past, when they would move just ahead of the great whales, courtiers to a monarch, gliding on invisible pressure waves created by the majestic progress of the largest animals that have ever lived.

Why dolphins jump clear of the water's surface is slightly more difficult to explain. There are several theories – some say it is to break the seal of water around their bodies, creating an instantly more efficient passage through the sea. Communication is thought to be another reason, the percussive thump of their bodies hitting the surface creating a message that is heard through many miles of ocean. The removal of parasites and leaping high to spot groups of fish in the distance have also been suggested, but I think there is another reason. It must surely be the sheer exuberance of breaking the shackles of the sea to soar through the air, if only for a moment, to twist that remarkable body and land in a detonation akin to a depth charge, or to knife back through the surface with barely a ripple.

I would visit these dolphins in the bay at Lochcarron many more times, when friends came to visit or simply when I had a day to myself. The reaction of anyone in the boat was always the same – pure excitement when the dolphins appeared, then quiet reflection on the way home. The leap of a dolphin is one of the great expressions of freedom in the animal kingdom, an act that binds not only them but us, one of the few sights on the planet guaranteed to make any human, regardless of race or creed, give a shout of pleasure and exhilaration. Throughout my six months on the west coast, I never tired of the dolphins in the tiny bay at Lochcarron, and – to my great surprise and pleasure – it seemed they never tired of me. Even on my final visit in the early grip of the autumn winds, they approached within a few metres again and again, before spiralling off into the dark waters of the loch, leaving vapour trails of tiny bubbles as an effervescent memory of their passing.

*

The sea was my constant companion at the cottage, always present and ever changing, a whispering portrait before me in the great window of the extension. It assaulted my senses, drawing my eye like a living thing, with the percussive heartbeat of the surf being the first sound I heard when I opened my eyes in the morning and the last thing that enveloped me at night. In the early months of my stay, it was a shimmering mirror, reflecting the dark mountains of Raasay, Rona and Skye, amplifying their bulk and splendour. Later in the summer, however, as the dark clouds gathered and the squalls cartwheeled up the channel, it became a dark, ominous presence, tearing at the land and hissing at the slope of the beach. The deepest water in Britain's coastal seas can be found just off the tip of Rona, plunging into abyssal gloom, the seabed hundreds of metres beneath home to organisms that live in perpetual darkness. There were times when the channel appeared to know of its echoing depths, becoming sinister and forbidding, a place where man should not venture. For a single week in the middle of summer the fishermen returned with tales of a fin whale feeding off the tip of the island, the second largest creature on earth lunging from deep water to feed at the surface. They spoke of the expanding bellows of its throat stretching and pulsing, sifting the sea a swimming pool at a time. I drove the RIB out to the spot where the whale had last been seen, bobbing over the deep water, feeling tiny and alone as I fruitlessly scanned the surface around me.

The ocean has always had an intense hold over me, even though I am certainly not from a seafaring family. My mother once confessed to feeling nauseous when looking at a picture of a ship in a storm, and once had to be led off a pier back to land, with pale skin, a clammy brow and wide eyes. My father was far more at home in the sky, spending much of his adult life cramming his angular frame into tiny cockpits to shriek over jungles and deserts with the RAF. I always felt that he had been slightly cheated by time, and really should have been lounging on the grass at Biggin Hill, having just sent the Hun packing before landing with only one wing remaining and his trousers on fire. Instead he ended up running a letting agency in Yeovil, staring wistfully into the sky and wishing his nickname was Binky.

My fascination with the sea emerged at a very early age. In the late 1960s, my family was posted to Malta, a blissful pseudo-colonial environment where we would spend inordinate amounts of time on the beach. With the unique confidence of a child, I would vanish beneath the waves, turning over stones in the shallows and scrambling buoyantly from rock to rock. My mother would sprint hysterically into the water when I disappeared from the surface for too long, only to find me pottering and bumbling along the seabed. As I grew older, I secretly hoped that I would discover an old sepia-tinted picture in the attic, showing someone in oilskins sporting a bushy white beard from which a corn-cob pipe emerged, with the words 'Take good care of the boy' scrawled on the back in seal's blood.

Alternatively, there was the possibility I would be whisked away by someone smelling strongly of kippers – one of the reasons I was always so attached to my granddad, now I come to think of it.

The sea around the British Isles is greatly underestimated by the modern generation, subjected as we are to relentless high-definition images of crystal-clear water, tropical reefs that seethe with life, and large toothy marine animals messily dispatching smaller ones. To ignore our own shallow waters is to overlook an impossibly complex environment of teeming riches.

As anyone who has subjected themselves to the festival of gurning that is swimming in Britain will tell you, the waters that bathe our shores are not warm. They are really quite chilly, creating a falsetto shriek on entry, and immediately turning nipples into passable impersonations of rugby studs. This is a very bad thing if you happen to be on holiday in Skegness, but a very good thing if you happen to be a marine animal. Cold water holds oxygen and nutrients far more efficiently than warm water, so the organisms around the coast of the British Isles bathe happily in cool seas. This is a reason why all the great fishing grounds around the world are in temperate seas.

The shallow seas around our shores are a mass of life that scuttles, crawls, slithers and sculls through nutrient-rich soup. There are an estimated 7,000 species of animals in our coastal waters, although this is a deliciously vague number – scientists always seem to get terribly shifty when asked this question, as the moment someone adds them all up, several new ones appear.

We are a maritime nation, and we rely heavily on our

fishing fleets. This is where one of the great dilemmas of modern legislation takes place, a battleground for not only the future of our seas but also one of our most traditional industries. What is undeniable is that the present level of exploitation cannot continue, with vacillating and dunderheaded bureaucrats passing toothless legislation as our fishing fleets trawl increasingly empty seas.

What is particularly mind-blowing is that the vast majority of fish caught off the British Isles are thrown back, and the majority of those will die a slow and lingering death, torn by the nets or ruptured within due to the sudden expansion of the swim bladders during the rapid ascent from depth. From 2002 to 2005, there were 186 million fish caught around the UK, a total of 72 million tonnes. Of this vast, glittering mass of protein, sixty-three per cent was thrown back, being either undersize or outside quotas produced by politicians. British waters – home to exquisite corals, the second largest fish in the sea, to dolphins, whales, and rocky reefs that shimmer and dart with life – have a grand total of three Marine Protected Areas, covering 0.001 per cent of our seabed. In the last twenty years, it is estimated that quotas have been set at thirty per cent higher than the sea can support, so it comes as no surprise that since the 1950s over sixty per cent of fish stocks in the UK have collapsed.

There is a glimmer of hope. The catchily named European Marine Strategy Framework Directive (EMSFD) of 2005 aims to achieve 'good environmental status' for member nations by 2021. The Marine Biological Association has identified 120 areas that need protecting, and the UK Marine Bill has called for fourteen to twenty per cent of our coastal waters to be offered some form of cover.

Balancing the needs of the fishermen with the parlous state of the marine environment is not an easy task. The present generation has the misfortune to be at the wrong end of a fishing spree that has lasted several thousand years; unless something is done, unless we stop shovelling gasping piles of young fish back into the sea or dredging the seabed for scallops, leaving behind a blasted, scarred underwater landscape, our children will inherit seas devoid of life.

I had spent many happy hours snorkelling in the shallows off Sand Bay. Over the course of the summer the ecosystem had changed, with the smaller fry that flitted nervously over the seaweed in the early months transformed into larger shoals of healthy survivors, the weak and the unwary stripped from their ranks. Prominent amongst these were the pollack, transformed from delicate will o' the wisps resplendent in rainbow colours, into dark, sleek predators, bossing their way along the reef. The brown bladderwrack thickened and spread, a dense carpet that itself twitched and flickered with smaller animals going about their business. Hermit crabs trundled busily from rock to rock, whilst the malevolent red eyes of swimming crabs peered out from the shadows beneath, armoured trolls in their caves. Brightly coloured anemones dotted the rocks like exquisite flowers, coming to life as the tide crept over them, delicate tentacles swaying in the current, seeking contact with the unwary, to enfold them lovingly in a slow embrace. I would occasionally see flat fish on the white sand, a creature that has the misfortune to be shaped like a mixture of a target and a frying pan. I speared one in the shallows, then felt guilty as I ate, the delicate white flesh tasting less like fish and more like sitting duck.

The arrival of the sand eels in June signalled a bonanza: gulls, gannets and puffins permanently patrolled the surface, whilst ravenous shoals of mackerel quartered the coast like wolf packs, ambushing the eels in explosive encounters that saw the surface of the water shimmer and heave at the fury of the frantic battle beneath. This was high summer, a call to arms for the larger predators, a festival of gluttony for seabirds, whales, dolphins and otters. The brightly coloured fishing vessels puttered relentlessly along the channel, quartering this way and that to reap their own harvest.

I was finally able to take Sue up on her offer of a dive, and rattled my way over the Bealach in the Land Rover. The dive was a delight. As we drifted over densely packed rock walls, several feet above the seabed like motherships over a moonscape, it seemed there was little moving in the mixed gravel and shells below. It was only when we touched down, under Sue's expert guidance, the true picture emerged. This was a mini Serengeti, with predators the size of my thumb chasing prey that blended chameleon-like into every dark crevice.

Peering closely at the jewelled surface of the gravel, I was somewhat startled to see a pair of tiny eyes staring back at me balefully. As I moved closer, the tiny stones behind the eyes began to move, a mini eruption that resulted in the appearance of a minute cuttlefish, its skin blotching and pulsing in alarm. This was the imaginatively named little cuttle, a perfect piece of biological engineering. It hovered briefly a few inches above the sea floor before whirring away into the blue of mid water.

Sue was busy a few feet away, and on drifting over I found her gently excavating a flame shell, a flamboyant series of orange tentacles around a delicate oval of a central carapace. This is one of those absurdly beautiful marine creatures that seemed to me almost garish, a bird of paradise of the reef. I could see Sue's eyes crinkle in delight at my obvious pleasure, before she released the shell, which swam away in a comical series of jerks, like an escaping set of chattering false teeth.

Moving up the rock wall, it seemed that every millimetre of rock face, was home to armoured, spiny, undulating life. Garish squat lobsters peered out of gloomy crevices, while gigantic edible crabs were crammed into the larger overhangs, claws held before them like pliers, beady eyes surveying us as we passed by.

As we moved into the final stages of the dive, we were into the waving rainforest of the kelp, home to the marauding pollack, loitering in ambush with a shifty air of being up to no good. Ballan wrasse bustled about closer to the sea floor, garish in their summer livery, the great opportunists, investigating anything as a potential food source, the Gemma and Doris of the underwater world. This particular wrasse has the extraordinary characteristic of starting life as a female, with certain individuals becoming males after several years of breeding, although in this species there is no change in the outward appearance of the fish itself. I always thought some sort of test using Hugh Grant films would be a good idea, but I satisfied myself by trying to spot the wrasse around me that looked mildly confused.

We sat in Sue's kitchen after the dive, spreading out charts and hatching plots for our next adventure.

'We should go here,' said Sue abruptly, jabbing a finger into the chart to emphasise her point. Her finger had come to settle on a headland off the northern tip of Skye, a cluster of rocks next to precipitous cliffs with the moniker 'The Maidens' written enticingly next to them. Beyond was the open sea; behind, the tallest cliffs in Britain; and beneath, a mysterious world of dark wall, overhang, cave and cavern.

'I don't know, Sue,' I said. 'It's tricky for me to get away from the bothy and the stock. I'm trying to be really serious about this, and can't really go off on wild goose chases. I'm fairly determined that if I'm going to do this properly I need to stay on site – my days of chasing the adventure are gone.' For emphasis I added, 'I really need to start being really disciplined about this. Sorry and all that, but I've got other priorities now.'

*

The full name of the point where we were anchored was MacLeod's Maidens, three pillars rising precipitously from the sea at Idrigill Point at the entrance to Loch Bracadale. There are a great many legends surrounding the Maidens, but the most likely story seems to be that they are named after the wife and two daughters of the fourteenth-century chief MacLeod, drowned in a shipwreck when returning from a battle where the chief himself was killed. The grandeur of the rocks seemed to reflect this story, forbidding and bleak, staring impassively into the vast, undulating plains of the Atlantic Ocean.

Our hosts for the day were Gordon and Aileen MacKay, the quintessential Scottish dive couple. He was a compact ball of muscle, every inch the teak-hard skipper, with an impressive scar etched on his lower jaw. When I tentatively asked about it, he merely barked the word 'Manta!' at me. This was surprising, as manta rays are known as the gentle giants of the ocean. He waited a moment – the classic dramatic pause – before adding 'Opel Manta, that is. Overcooked it on a corner, kind of got away from me.' He gave a bark of laughter, then returned to preparing the boat for the dive, another victim in the bag.

Aileen was a bubbly presence, bustling about the boat like a benign mother hen. She was a very well-qualified diver in her own right, but for this trip she had the key role of providing me with a steady supply of flapjacks, home-cooked weighty slabs of oats bound together by honey from the heather. After a while my compliments dried up, and I would simply gaze at her with cow eyes, sending her below decks for yet another golden ingot of energy, which I would stuff into my slack jaw like a stoker feeding a raging furnace.

In response to our request for a very exciting dive, Gordon had immediately suggested Conger Crevice. This, he said, was a medusa's lair of twisting slate-grey bodies, a viper's nest of gigantic eels alongside clicking hordes of lobsters and crayfish. It sounded irresistible, and in no time at all we were kitting up alongside the dark rocks that led to the overhang that was the home of the eels.

The conger eel is one of the giants of the reef and wreck, something out of a child's fantasy, a huge, snakelike body curled in its lair, peering at the world through dark eyes and venturing out on the hunt in the hours of darkness. Its life cycle is remarkable, spawning in the deep water of the mid Atlantic, a journey of several thousand miles to a birth place that is an icy cold, pitch black plain – far beyond the reach of light and life as we know it. After breeding, it will die, with the new generation heading towards the shores of Britain to repeat the cycle.

The congers have a reputation for ferocity that is based entirely on its thrashing death throes on the deck of many a fishing vessel. They react in much the same way you or I would if dragged from our homes and clubbed to death, being somewhat miffed by the process. In their element, they are cautious, graceful animals, retreating at the approach of divers and having to be patiently coaxed from their lair. This belies their size and power – the British record for a conger eel is 133lbs (60.33kg), although a 350lb (158.76kg) leviathan was caught off Iceland in a commercial trawl. These big congers are hugely impressive – over two metres long, with dark heads the width of a grown man's waist. I had seen huge congers on shipwrecks in the past, and had always had the impression that they simply wanted me to pass on by, slinking back into the shadows to continue to feed and grow, waiting for their final journey into the echoing depths.

Sue and I completed our checks, lumpen and angular in the various accoutrements required for exploring the sea around the British Isles. The main problem with any dive in Britain is the cold – the waters around us were a frigid 13 Celsius, even at the end of the summer. Water conducts heat twenty-five times faster than air; it's not a case of the heat seeping out, it is more that the cold creeps in, threading and weaving its way into your bones, an invasive chill that leaves one fumbling and stupefied.

I therefore wore my drysuit over an undersuit (the delightfully named 'woolly bear') – essentially a giant romper suit consisting of a many-togged duvet with arms and legs.

The drawback was that it made you very buoyant indeed, so great lumps of lead were required to get you below the surface and keep you there.

Pink-faced and perspiring, we waddled our way to the side of the boat like morbidly obese penguins. We took our seats with much clanking and hissing, then rolled backwards into the water, hitting the surface with a geyser of spray and foam, multicoloured depth charges exploding through the surface and into another world.

The rock face soon appeared before us, a mass of kelp fronds waving in the current, creating shadows which flickered and flashed over the crenulated rock face beneath. Following the wall down into the deep green below, the jade of mid water was replaced by the darker tones of the seabed. Where the sea floor met the bottom of the cliff, there was a dark overhang, a suitably atmospheric home for the sinuous eels within.

Sue was already on her belly, worming her way into the dark crevice with legs wriggling behind her. As I joined her on the sea floor, my torch beam played across a living fresco of shellfish, crustaceans and sponges on the surface; reds, yellows and cobalt blues ignited into a tapestry of conflicting hues, as vivid as any coral reef.

By this time, I had been joined by a cuckoo wrasse, a bold reef fish that bustled and fussed around me as I moved down the wall. This was a male in all his courting finery, a riot of whorls, lines and spots, a love letter writ large on his shimmering flanks. He would remain with me throughout the dive, moving ahead and then charging back as if to check on my welfare, a conscientious tour guide resplendent in his garish livery.

I tentatively moved my head down to peer under the overhang, and immediately made out the unmistakable form of a conger. The massive, slate-grey body lay full length at the very back of the overhang, waiting for the coming of the night when the hunt would commence. Sue was already taking photographs, the strobes of her camera illuminating the scene like cracks of lightning from some second-rate horror film. She turned to look at me, eyes wide, a smile creasing the corners of her mouth amid a stream of bubbles.

I moved further along the overhang, and there before me was another gigantic eel, this one turning towards me to peer through dark eyes at the strange creature before it. I lifted my own camera and began to take a series of shots as the eel moved and pulsated before me.

It was only after several shots that I realised that the conger was pulsating very much in my direction, growing in my viewfinder with each flash of the strobe. It had no ill intentions, simply wanting to make its way to somewhere dark and secure, away from this noisome interloper that had dropped in from the skies, all flashing lights and thundering bubbles.

The problem was that I was firmly wedged in the direct path of the eel, and when a conger decides that it's going somewhere, it generally gets there. This may require the removal of certain boggle-eyed, wildly reversing creatures in its path, but as most things get out of the way of congers, this one couldn't see any reason why I would do anything other than scuttle backwards at great speed.

Surprising both myself and my brightly coloured wrasse guide, I managed to reverse very successfully, emerging from the overhang like a cork from a bottle, in a hail of bubbles, cycling knees, and trailing hoses. The conger followed me out, gliding towards a distant reef, an airborne anaconda above an undulating rainforest.

The rest of the dive passed with considerably less drama, a glorious voyage along the steep wall of the undersea cliff, our depth gradually decreasing as the seabed angled upwards, but the scene around us remaining as busy as ever. Greeting us on our final stop - a drift just below the surface before our last ascent - was the gently pulsing form of a lion's mane jellyfish. These delicate, beautiful creatures are as ephemeral as the water itself, and yet pack a deadly punch in their tentacles, millions of tiny harpoons called pneumatocysts primed to fire at the touch of any passing fish. They drift through the sea, orbiting the shallows like heaving motherships - gossamer, flimsy vessels trailing death in their wake.

As we puttered home at the end of the day, diving kit packed away, shoulders hunched into a warm fleece and a cup of hot tea chasing the cold from within, I reflected on the riches of the seas that lap our shores. The cliffs we passed were still touched with the light of the setting sun, illuminating the squabbling, shrieking hordes of gannets and shags that perched on their ledges. Much as this was a scene of boisterous life, for me the real excitement began at the surf line, a crackling border between worlds, beyond which lurked caverns, forests, giants and unimaginable adventures.

7

The Crofter's Way

Certain issues dogged me as much as the crofters in whose footsteps I hesitantly followed. It was my own fault that I initially failed to make the transition to a life without electricity – something to which we are all subconsciously geared in the western world. Our first breath is taken in the glare of electric lights, our bodies warmed by the whirr of distant turbines, our internal health immediately monitored by a series of beeping, whirring panels and boxes. From that point on, we are surrounded by electricity, it is a part of our lives that hums quietly in the background, serving us but in the process making us beholden to it. Without it, we become instantly helpless, vulnerable infants. Electricity is natural selection's way of softening us up before polishing us off. Even in the idyllic wilderness of the west coast I could not survive without it, and ended up erecting a wind turbine that hummed gently on the hill behind the cottage.

The electricity it provided was enough to light the bothy and power my radio, both of which would – of course – have been truly miraculous to any crofter only two generations before.

I have a faintly apocalyptic theory about all this. I feel that we are enjoying a gilded existence unprecedented in the history of mankind. Warmth is the turning of a dial, we are bathed in light at the flick of a switch, and communicate on a global scale by tapping a keyboard. This cannot last, as we all know at some level. I fear that we're all heading gleefully to oblivion in a headlong free-for-all. In a few generations, we'll all be crouched in caves, making a fire in the sooty box of what was once a television, and telling the grandchildren stories about *Big Brother* and other abominations.

With such a gloomy outlook on the future of mankind, the acquisition of some basic skills during my time in the bothy seemed to be a good idea. Foremost amongst these was the ability to store food for protracted periods. Back in Bristol, I would simply buy on a massive scale and store food in the freezer, and yet in the bothy this mindset led to vast amounts of food being thrown away. I found this completely unacceptable, a criminal squandering of precious calories.

Having food lying around also led me to become slightly more closely acquainted with some of the local wildlife than I would have liked. Within a couple of weeks of moving into the bothy, I was joined one evening by a mouse in the extension. If this was a shock for me, it was positively life-altering for Reuben.

The evening was a typically tranquil affair, one of a relentless series of golden dusks that will be my lingering memory of the early weeks in the bothy. There was not a whisper of wind, the midges had yet to arrive, and the sea was gently caressing the rocks of the shore only metres away. I was writing in the extension, harvesting the remaining light before turning in for the night. Reuben was curled up by the door, a dark mass twitching and snuffling with the memories of the day.

As I wrote, I became aware of a movement out of the corner of my eye. Glancing down, I saw a tiny mouse scuttle busily out from under the sofa. He paused to look briefly up at me, conducted a brief face wash with tiny paws, then continued on his way. He was very cute indeed, with soft brown fur, tiny scoops of ears, and shining black buttons of eyes that glanced around with considerable enthusiasm and interest at his new surroundings. I always feel that mice seem to be humming and muttering as they investigate a room, and this chap was no exception, flitting from side to side, all whiskers and scuttling energy.

His investigations inevitably led him to the great dark mountain that was Reuben. I watched as he darted ever closer to the dog, with one final energetic scuttle stopping only inches from his dark muzzle. Here the mouse paused, sat back on glossy haunches with his tiny paws held primly to his chest, and surveyed the colossus before him.

An inner alarm sounded within Reuben, who became peripherally aware that he was being watched. Without moving a muscle, he lazily opened a slit of an eye, and focused on the mouse as it sat inches from his nose.

I'm guessing that there was a considerable problem of scale and perspective. By now the mouse pretty much filled Reuben's field of view, and must have looked like some sort of freak rodent the size of a Volkswagen Beetle. The dog's reaction surprised me slightly, and surprised the mouse a great deal. Reuben leapt in a single motion to his feet, simultaneously letting out a blend of a yelp and a bark, his eyes wide and ears flat, lips drawn back and his face fixed in a rictus of alarm. His entire body was trying very hard to move away from the gigantic nightmare mouse before him, creating an explosive cringing leap that in turn caused his large hairy backside to crash into the wooden door of the extension, which rattled alarmingly on its hinges. The impact deposited him back on the planks of the wooden floor, his gaze fixing on the rapidly retreating form of the mouse, which was just vanishing back under the sofa. This caused him to reconsider his initial reaction, and glance back at me with something approaching embarrassment.

Delighted as I was with having seen a wild mammal on the west coast, I realised very swiftly that having food lying around going off would not do at all, and resolved to build a smoker.

Keith Jackson was my obvious first port of call, and very shortly afterwards I was wandering around in the heather on the hill, mobile phone held aloft, bellowing into it as he tried to make sense of what I was saying.

'Ah, a smoker. Got you,' he said finally. 'Monty, it's a piece of cake. In fact, I built my first one when I was ten . . .' No surprise there, I thought. 'You basically need a means of directing smoke into an enclosed space,' he went on. 'A fire of course, and a means of ensuring that the temperature doesn't rise too high in the container. You're looking to smoke it, not cook it.'

He chatted for a little longer, offering tips and advice based on his own efforts over the years, before signing off.

'Basically, Monty, as with most things, it's simply a case of getting on with it. Have a go, I'll see if I can come over in the next couple of weeks to check it out, and don't set fire to yourself or the bothy. Oh, and if it all goes wrong, I'm absolutely nothing to do with it,' he finished with a chuckle. Wishing me good luck, he hung up.

I woke the next morning very excited and ready to take one step closer to self-sufficiency. This was a relatively light-hearted project for me – I could have kept driving over to the village shop on a daily basis and survived quite happily – but for the crofters of old it was very serious stuff indeed. The ability to reap the whirlwind of the summer harvest on land and from the sea was only valid if the food could be stored for the long, lean, bitter months of winter. As such, salting, smoking, drying and pickling were all important parts of Highland life.

I felt the normal tide of helplessness rush over me as I looked at the pile of wood I had assembled at the side of the bothy. Glancing down at the hammer in my hand and the bulging bag of large nails (I had adopted the classic Monty Halls approach to DIY for this project), I decided the best thing to do, as Keith suggested, was simply get stuck in.

Several hours later, I had a bandaged thumb, a blood-spattered sleeve, gritted teeth, and – miracle of miracles – a passable imitation of a smoker.

My smoker – of which I was inordinately proud – consisted of a shallow pit about 180 centimetres long, 60 centimetres wide and 30 centimetres deep. Over this were two pieces of chipboard, hammered together by several thousand nails, with a large piece of flat metal at one end that would act as a shield for the fire beneath. The other end of the smoker was sealed except for a slight gap to allow smoke to be drawn through. On the inner apex hung several hooks, designed to hold the meat and fish that I would be smoking.

There are essentially two types of smoking – cold and hot. The latter cooks the meat or fish, and requires a temperature range of 165 to 185F for that perfect, moist, smoky flavour. Cold smoking requires that the temperature within the chamber does not rise above 100F. Should it rise above this magical figure, it essentially becomes a very inefficient cooker, with temperatures fluctuating within and all manner of virulent pathogens appearing in the meat or fish being smoked. These will make you very ill indeed, so it is crucial that the temperature in a cold smoker is within a certain range. It is playing with fire in the truest sense.

All sorts of things can be used to create the fire for smoking. In Europe it has always tended to be the hardwoods – oak, beech, and alder. In the USA it tends to be hickory and maple, although pretty much anything that produces non-toxic smoke is worth a bash – corncobs, peat, old tea, sugar, rice – all are used to create various delicious smoked concoctions around the world.

Traditionally, cold smoking has also been combined with salting, as the smoke needs to penetrate deep into the meat to aid preservation. I duly spent the afternoon rubbing salt into various meat joints, before suspending them within the smoker. The final stage was building the fire, on my hands and knees coughing, spluttering, red-eyed and annoyed, before it began to billow smoke like it meant it. The fire needed to burn for three days – which is a bloody long time in anyone's books – meaning a series of nocturnal tramps out to the smoker muttering darkly. Finally, the meat could be removed, tanned to the colour of old leather, artificially aged and mellowed.

I prepared my first meal that evening – smoked lamb with a dark jus of raspberries – aware that this would either prove a delightful culinary experience and reinforcement of my survival credentials, or it would result in my premature death in a noxious miasma of spreading bodily fluids. Thankfully, the meat was delicious; warm, fuggy and deeply smoky, yet offset by the sweetness of the sauce. The dark colours combined on the plate with the rich brown of the lamb and the arterial purple of the sauce, glistening in the candlelight, the sweet smell dancing on my taste buds and seeping into my sinuses.

I went to bed that night thinking that if this was going to be my last meal, it had been a pretty good one.

I awoke the next morning feeling a new era had dawned. The next few weeks were spent smoking everything I could lay my hands on, happy in the knowledge that I was one step closer to realising the elusive dream of self-sufficiency.

*

My time in Applecross had – up to this stage – been truly idyllic. Over the previous three months, I had built the bothy into something approaching a home and had even welcomed my girlfriend Antje for an extended visit, adding some finer touches in anticipation of her arrival. As with so many aspects of my new life in Scotland, I was doing something very basic – carving out a life for my family coming up to join me. It had all been undeniably hard work, at times frustrating, but never a day went by without me sitting for a moment looking across the waters of the channel and feeling a little of the magic of Applecross creep further into my soul.

There is, however, another side to crofting. Up to this point, I had essentially been running a small petting zoo, leaning contentedly on the rough walls of the pig enclosure with a grass stalk twirling between my lips, crouched among the chickens whilst hand-feeding them corn, or watching the wary movements of the sheep through the deer fence. A call from Keith in early July snapped me out of my reverie, his voice as jolly as ever on the end of the phone.

'Monty, I think it's time I taught you a little butchery. It's no good you having those sheep eating all your feed for nothing – how about we assist one of them on their journey into the fridge?'

I was taken aback at the thought of putting down one of my animals, particularly when my entire focus to this point had been on keeping the things alive.

'I . . . erm . . . ahhh.'

'I'll see you next week,' Keith laughed. 'Don't worry, it's all part of the big picture of crofting, and I'll talk you through every stage. See you on Wednesday, my boy.'

Keith had spent his entire adult life gunning down doe-eyed, moist-nosed mammals and had thoroughly mastered the ethics and mechanics of the clean kill. I, on the other hand, had never killed anything larger than a mackerel. In one notorious incident from my Marines days, I had been given a rabbit to slaughter on a survival exercise. I made the mistake of hanging on to it for a day or so, giving it a name and generally bonding with it. Cometh the hour, as Thumper stared up at me with twitching whiskers and trembling lower lip, I couldn't do it and had to hand him over to a Glaswegian Marine on the understanding that I would prepare his vegetables if he'd polish off the rabbit. He duly took Thumper off and murdered him behind a nearby hill whilst I fine-chopped and diced his turnip in return.

Keith and Rachael turned up on the day in question, carrying with them a fearsome set of tools, glinting dully in the morning light. Jess and Tristan were with them, playing happily in front of the bothy as Keith talked me through the kill. Once the rifle was out of the bag, his mood changed, becoming more focused and intense.

'Monty, it's a case of getting in the right position and doing the job quickly and cleanly,' he told me. 'This isn't a snap shot at a running animal hundreds of metres away – you'll shoot it in the front of the head from a range of about six inches, and you'll do it when I tell you to. You've raised this animal well, and it's up to you to give it a decent, quick death.'

Rachael was, as ever, more sympathetic, offering quiet words of support before I headed grimly into the field. Keith walked behind carrying a bucket of feed to draw the sheep in.

It seemed to me that everyone and everything had picked up on the mood of the moment. The wind seemed to raise a notch, with the dark clouds that had been present all morning crowding in. The waves crashed onto the beach in messy, small breakers, and the bracken whipped and twitched against the dry-stone walls. I turned my collar up against the sudden chill, chin to my chest and shoulders hunched. Reuben had gone unusually quiet, and was lying with his back to me, head down on the wet grass. Rachael ushered the children into the house, and we approached the herd as they rested in the thick bracken in the far end of the pen.

There is a very good reason why Soay sheep have prospered in this hard land, remaining resolutely fixed in the wilds of Scotland as other species have come and gone. As well as being remarkably hardy, adaptable, nimble and bright, they also possess that most valuable resource of any animal – the indefinable sense that something is not quite right. As Keith and I approached, with him shaking the bucket and me attempting to look like I was out for a midday stroll, an effect ruined by the fact that I was carrying a loaded rifle and looking utterly miserable, the sheep stood stock-still. The ram was particularly wary, and refused to approach even when the feed was placed on the ground. In an extraordinary moment of animal perception, he even stayed well back as the other sheep finally began to move towards us. This was in complete contrast to his normal feeding behaviour, which tended to consist of advancing horns first at considerable speed, butting aside anything that got in his way. He walked back and forth along a slight grassy rise in the field, always out of range of the kill, always watching me, tossing his head and starting at every movement.

After nearly an hour – one of the most excruciating of my life – he was at last within a few feet of me. He slowly dropped his gaze from mine, and lowered that armoured head to nibble at the feed spread at my feet. Keith nodded silently to me, and I stepped forward, placed the rifle against his woolly forehead and, feeling I was betraying every piece of trust any animal had ever placed in me, pulled the trigger.

He dropped like a stone. Keith moved forward swiftly – patting me once on the back as he passed – and threw a length of rope beside the twitching corpse. He turned to me, the wind whisking the smoke away from the ragged cigarette he was smoking, and gave a half-smile.

'Well done, Monty,' he said. 'It's hard, I know, but if it was easy it would mean you didn't have farmer's blood in your veins. The fact that you care so much for this animal means that maybe we'll make a crofter out of you yet. Now let's get this fellow hung up and bled, and then we can start the butchery.'

The first crucial step was to bleed the ram, which had to take place immediately. Dragging the still-warm body to the fence, I strung up the carcass by the heels, and under Keith's direction slipped a knife into the soft, pulsing skin of the neck. Immediately there was a ruby geyser of arterial blood, spattering into the bowl Keith had placed on the ground beneath.

When the flow had subsided, he handed me a whisk with a grin. 'Right, Monty, get whisking. The blood can't be allowed to coagulate.'

I added salt to the blood, and began the macabre task of whisking it. This prevents coagulation before it cools, and soon I was spattered with red spots, blinking through a gory mask and fighting the urge to charge off to the pub to demand a green salad.

Keith danced around me throughout. Government legislation surrounding killing livestock on your own property meant that he was not allowed to touch the carcass at any time, and as such he was virtually exploding with frustration.

'Come on,' he barked, 'whisk the stuff like you mean it. Man like you, living alone for so long, should have an incredibly strong right arm. Now get stuck in.'

I had spent the last few months being rather scared of Keith, being quietly convinced that if I upset him he'd turn me into some sausages and a novelty hearth rug. I whisked like a man possessed.

'OK, that's good – now, now. Come on, you muppet, no time to waste.'

I mixed the blood with pepper, oatmeal, suet and then stuffed it into skins. Keith had already built a fire that roared like a blast furnace, and bustled me across to the pot that boiled and bubbled on top.

'In there, in there,' he said, pointing unnecessarily to the steaming interior as he danced from foot to foot. I dropped the sausages in gingerly.

He bounded over to the cutting table that had been laid out prior to the kill. 'Right, now we're going to create the meat joints. Come here and I'll show you how and where to cut.'

He had laid out a gleaming array of tools, all ferocious blades and wicked serrated edges. Although I had not enjoyed the moment of the kill at all, my eyes lit up at this buffet of decapitation. This, I thought, was going to be splendid.

With Keith's finger darting in and out, pointing out the precise places to hack and saw, I attacked the carcass. The finger strayed dangerously close during some of my more flamboyant movements, particularly the Nadal-style over-arm smash I employed to split the pelvis. Some of the latter cuts even included a run-up that a West Indian fast bowler would have been proud of. By now I was a wild-eyed savage, and may even have produced a guttural grunt or two in moments of high emotion. In no time at all, the ram was in its constituent parts.

These were salted, then vacuum-packed in a wondrous machine Keith had brought with him. The meat would be buried in a deep hole outside the bothy, being dug at that very moment by Tristan, ably assisted by Reuben. Here they would lie for five days, the salt seeping into the flesh as a preservative, before being stored.

The heart and lungs had been bubbling away on the stove throughout the cutting process, and now we mixed these with oatmeal, suet, onion, salt, cayenne pepper, nutmeg, herbs and milk to create haggis. When I had created the mix, Keith leaned over to inspect it.

'Get your hand in there, Monty, and try some.'

It was absolutely delicious.

'Mountain food, that is,' Keith told me. 'If there had been a bit more of that around when you buggers came north over the border we'd have given you the pasting you deserved.' He laughed uproariously.

The haggis duly disappeared into the bubbling maelstrom of the pot, before I turned my attention to the trimmings from the butchery.

'Nothing wasted, Monty – you're a crofter now,' said Keith, who appeared to all intents and purposes to have gone temporarily insane.

The trimmings were minced and combined with salt, pepper, herbs and suet, then fed through a hand-cranked sausage machine. I don't care how old you are and how serious your scientific or organic credentials may be, when sausage comes out of the nozzle of a hand cranked machine it looks a lot like a willy or possibly a poo, and is very funny indeed. Keith had seen it a million times before, and even he laughed.

By the end I looked like a music-hall villain, spattered in bits of ram. Pots bubbled and boiled merrily away in the background, sausages hung from the rafters like turgid fruit, the air was redolent with the smell of meat and fat, and Reuben was doing laps of the bothy staring at the sausages in the roof with tongue lolling.

When the last piece of meat had been buried and the final sausage placed in the pot, Keith wandered over and shook my hand. As we chatted, it struck me what a remarkable man he was, his tremendous enthusiasm and energy backed by a vast array of skills. Should there ever be a nuclear holocaust, only rats, cockroaches and Keith would survive.

'There we go, Monty,' he said, 'that's the crofter's way. Everything used but the bleat.'

8

Wild and Free

The weather finally began to settle in early July, just when I had reached a point of mild desperation. I was aware that the clock was ticking for me, the moment when I would have to hand the bothy back to the Applecross Trust now being only ten weeks away. I simply couldn't bear to think of the long journey back down the length of Britain to my home in Bristol. Climbing up the twisting snake of the Bealach na Ba for the final time would, I knew, be a moment of real emotion.

I needed to make the most of the last two months in this magical place and the lull in the weather created a window of opportunity, one I decided to exploit fully.

All of the best leads and tips invariably came about halfway down the second pint in the Applecross Inn, and the previous Friday had been no exception. I was sitting close to the fire with Mike, Linda and Sam, and had mentioned that I was keen to see the larger visitors in the channel next to the bothy, something that had always been a powerful motivating factor in choosing Scotland as the location for my great escape.

'Do you know,' I said, warmed by the glow of the fire and mellowed by good company, 'the largest mammal I've seen so far in the sea off the bothy is Reuben.'

We all glanced down at Reuben, who was curled at my feet. Admittedly he was a very large mammal indeed, but not quite the magnificent animal encounter I'd had in mind when I had first found the bothy.

'Well, Monty, it seems to me that you're missing the blindingly obvious,' said Mike, who knew these waters as well as anyone. 'You sit every day looking straight at Raasay and Rona – if you want to see some big marine animals up close and personal, why not climb into your boat and head to the northern end of Rona. There's a huge grey seal colony there – you'll have some lovely encounters for sure.'

This seemed the perfect opportunity for one of my first trips with Antje, who had come to stay with me at the end of June, and the next morning we loaded up the RIB for the long haul out to Rona. Living in a windy bothy on the west coast was not as much of a shock for Antje as it would have been for others; a field biologist by trade, she travelled the world as a specialist in penguins, generally getting down and dirty wherever bird colonies are present. Compared to that, perhaps our trip was tame, but this was still no idle potter in the boat – the GPS showed a distance of 23 kilometres, and the colony itself looked straight onto the open Atlantic. Lose power when off the colony, and the result would either be a call over the radio to be rescued by a tittering local fisherman, or (more preferably) a slow drift into the Atlantic, followed by dehydration, a certain amount of urine drinking, attempts to eat the kapok stuffing of our life jackets, waving the tattered remains of our underwear at passing jumbo jets, then a slow and lingering death.

The passage up the channel was glorious, with the forecast proving accurate for once. The locals told me that the most reliable weather forecast could be found between the curtains every morning but, since I lacked the knowledge to accurately assess the impending conditions, I persisted in trusting the sonorous tones of the shipping news. This in itself was a delightful development, marking the profound change in my life over the last few months. The shipping forecast had always had the whiff of romance, with the names themselves – Dogger, Fastnet, North Utsira, South Utsira – speaking of massive grey rollers and fishing boats battling home through explosions of spray. I had always imagined fishermen crouched over tinny radios, exchanging ashen glances due to the impending arrival of spiralling low-pressure systems. It all seemed impossibly romantic, wild and admirable, a million miles from my centrally heated world in Bristol. Now it meant so much more, a clipped voice telling me that it would be safe for me to put to sea, or warning me to stay put in the bothy and put the kettle on.

That morning it had talked of fresh north-westerlies, which I could now feel in my face as the RIB skipped and danced from crest to trough, surfing and sliding over gentle swells. The stark hills of Raasay passed to our left, rugged shores unsullied by the hand of man. Periodically we would slow to creep towards a puffin, sitting plumply on the surface. Puffins are inherently comical birds, with stubby wings that have to flap an incredible three to four hundred beats a minute to launch their absurdly un-aerodynamic bodies into the air. As such, their take-offs from the surface of the sea seem to involve a certain amount of crashing into wave crests and bouncing along on what looks like a pot belly. This was the midst of the breeding season for the puffins, and we saw numerous birds heading towards their colonies on distant cliffs. The fact that the young puffins, noisily awaiting the results of the adult birds' fishing forays, are called pufflings reassures me that perhaps science isn't quite as dry and humourless as we sometimes think.

Creeping west through the channel between Raasay and Rona, we immediately encountered the massive glassy swells of the open ocean as we rode up the exposed west coast. Millions of tonnes of waters slid beneath our keel every few seconds, a gentle roller coaster guiding us towards the seal colony. What this coastline must be like in the midst of the fury of a winter storm beggars imagination – a scene of relentless power, and unfettered ferocity. The suppressed energy beneath us was palpable, held in place by the film of the surface, a coiled leviathan ready to turn in a moment and dash our flimsy craft against the rock walls if the mood took it.

With some relief, we turned into the inlet where the seals had settled into a large, odorous, honking colony. The arrival of a bright orange RIB seemed to take everyone by surprise, and immediately the seals started to slide and hump their overstuffed bodies into the water, changing in an instant from plump, waddling matrons into twisting, darting missiles leaving vapour trails of shining bubbles.

The inlet itself was very beautiful, with low, craggy rock shelves providing the perfect location for the seals to haul their rotund bodies out of the water. The main entrance to the sea was narrow, only fifty metres wide, and yet the curve of the rocky shore at the back of the colony measured several hundred metres. Past the shore, grassy slopes led into the island's interior, steep enough to provide shelter from the cold easterly winds. The air was heavy with a dense, oily, fishy odour.

We dropped anchor and began to prepare our snorkelling gear. Observing seals from the shore is one thing – and very rewarding it is too in the right circumstances. To really appreciate them, however, you must share their element, the blue amphitheatre through which they soar beneath the silver sky of the water's surface. Here their confidence is heightened, and when this is combined with their natural curiosity it can lead to truly magical encounters.

The grey seal is something of an anomaly in the animal world, breeding in the autumn and thus requiring their pups to brave the worst of the winter storms. This is only the beginning of their quirky breeding attributes. The young emerge snowy white – a throwback to when Britain was covered in ice. As the seals have had no significant natural predators since the ice retreated, there has been very little point in reverting to the duller grey of other seal species pups, and as such the colonies during breeding season are dotted with doe-eyed balls of white fluff looking completely incongruous against the dark rocks of the British coast. If anything, being conspicuous may help their chances of survival, as huge numbers are crushed by charging bull seals controlling their harems, 300kg of galloping lard bearing down on them. The seal pups appear in great numbers in a short period, a breeding strategy from the females designed to swamp any predators with potential prey items. This is once again an evolutionary throwback, as their natural predators have moved north with the receding ice, although orcas may take the odd adult and possibly a transient large shark may take a young animal.

There has been very little need for natural predators as man has been doing a very good job indeed of eradicating the grey seal. Before the Grey Seal Act of 1914, numbers were down to as few as 500; they have now rebounded to a healthy 124,000. Even with this larger population, there are considerable threats: mortality amongst the young is estimated to be as high as sixty per cent for the first year of life, and a mysterious virus has been wiping out huge numbers around the British coast. Life in the big wide world is no easy matter for a young seal.

By now the boat was surrounded by inquisitive heads poking through the water's surface, with one youngster particularly curious. Time and again, he would edge closer and closer to the boat, wide eyes fixed on the strange figures bustling within. He seemed to be playing a strange game of bluff, creeping up as we pretended not to see him, and then diving in an explosion of spray the moment we glanced in his direction, appearing seconds later the other side of the boat to commence his stealthy approach. What he was going to do when he actually got to the boat was anyone's guess, and something that I'm fairly sure he hadn't quite figured out himself yet, but he seemed to me the sort of maverick youth who would deal with that particular eventuality when it finally happened.

By now, Antje and I were ready to slip into the chill sea, and did so with barely a ripple, to scull slowly away into the open water. Hanging at the surface, we scanned the seabed beneath us, patches of white sand interspersed with dark, mysterious areas of kelp. Within seconds, I heard a squeak from beside me, the unmistakable sound of an excited penguin biologist saying things very rapidly through a snorkel. I followed her gaze, and there beneath us was the young seal, gliding over the seabed on its back, peering directly up at us as he passed beneath.

To watch a seal in the water is to see an animal that is the master of its element. The seal cavorts in three dimensions, freed from the shackles of gravity, a twisting, turning, undulating expression of complete abandon. The speed of movement is breathtaking, appearing before you one moment then behind you the next, confidence growing as it observes your clumsy movements in response to its own. Seals are diving machines, perfectly adapted without and within. The grey seal can dive to 270 metres, storing oxygen in haemoglobin-rich blood, becoming in effect a plump diving cylinder. Once this oxygen supply is depleted, most of the organs can continue to operate, although recovery from these longer dives can take some time. The pulse rate slows on the dive, the body cavities flatten to avoid the risk of decompression sickness, and the seal becomes a voyager soaring through the depths, equipped with the ultimate physiological equipment evolution can provide. What they made of my own efforts I have no idea, although the young seal darted forward during one of my clumsier dives to nip the end of one of my diving fins, the seal equivalent of ruffling my hair and pulling my shorts down.

As we rolled and danced with the seals in the clear waters of the bay, the minutes became hours and the sun tracked across the sky. Glancing at my watch, I noticed that it was well beyond the time when we should have been heading for home. Retrieving Antje, who by now was waving a large frond of kelp and making gurgling noises at the young seal through her snorkel, we reluctantly made our way back to the boat. Clambering aboard we instantly felt the chill of the afternoon breeze, clumsily changing with fingers made blue with cold. Even as we shrugged into our warm clothes, the young seal continued to cavort around the hull, ignoring the impatient coughs of its mother on the nearby rocks. The echo of his splashing and rolling was amplified by the rugged walls of the bay, a farewell salute to send us on our way as we nosed into the wild waters off the tip of the island to begin the long journey home.

*

Although my time in the bothy was turning out to be everything I had hoped it would be, with every day slightly different and the landscape subtly altering around me as the summer rolled on, there was still one major ambition that remained unfulfilled.

I had already had two fleeting, heart-stopping glimpses of an otter on the shore next to the bothy. The first was in the most bizarre circumstances during the latter stages of the build. As we hammered, hacked and sawed, creating a colossal din that resonated around the entire bay, someone had shouted and pointed into the middle distance. There, between where we were noisily building and where Reuben played in the sand, was an otter. Flowing over the rocks like a glob of mercury, it glanced casually in our direction, before continuing on its way, heading inland towards, I imagined, the burn that babbled away at the head of the bay.

Several weeks later, shortly after I had moved into the bothy, I saw another one. This time, it was diving in the clear waters off the headland, an unmistakable roll of a slick back arching beneath the surface, leaving only a small splash and a spreading ring of shining ripples. I had scanned the waters around the headland for an hour afterwards, desperate for another glimpse, but the otter had moved on, doubtless busy with the hunt. I spent every evening for the next week crouched behind a rock, long lens pointing hopefully at the stretch of water where the otter had briefly appeared. Although this had been only a momentary sighting, it had burned itself on my memory, a stark negative of a lean body sinuously rolling through the surface into the depths beneath – quite the most languid, beautiful movement I had seen in any animal so far on my west coast sojourn.

The otter seems an animal synonymous with freedom, and yet with a certain irony they have been anything but free over the last century, having been hunted to near extinction until granted full protection in 1978. Although the otters along the west coast of Scotland seem to favour the sea, they are not a purely marine species. They and the river otter are one and the same – *Lutra lutra*, the Eurasian otter – the link between the two lying in the fact that even the otters of the west coast need a fresh water source nearby, in order to groom and to drink. The Eurasian otter can be found from Britain to Japan, and from North Africa through to Finland, but the population on the west coast of Scotland is one of the most significant in Europe, a stronghold for the species, even when they were being eradicated in Wales and England.

Early summer is a busy time for them, as litters of up to five young are born in spring. The cubs are completely dependent on the mother for the first six months of their lives, remaining close to the holt – a hole dug between tree roots, under boulders, or into earth banks. The mother has a full-time job feeding the litter and is an opportunistic hunter, seeking out fish, crabs, urchins, small mammals and, in times of hardship, even insects. On the west coast they have a particular penchant for butterfish, a slippery, eel-like fish that hides under rocks and in crevices in the shallows.

I mentioned to Kayak Mike my considerable frustration at not seeing more of the otter. As ever, he had an immediate solution.

'We have a beach locally that we try to keep to ourselves,' he said. 'Buy me another pint, and I may consider telling you about it.'

Moments later, top lip covered in white froth, Mike leaned forward conspiratorially. 'There's a tiny beach north of where you're based. I'm pretty sure that this is the centre of an otter's territory – head up there just before dusk, sit quietly, and you may get lucky.'

He rummaged in his rucksack and produced a creased map, smoothing it onto the table. He pointed to a rocky bay, a tiny dent in the craggy coastline.

'There we go, Monty,' he said. 'That's where you'll find your otters.'

It looked wonderful – a smugglers' cove with a river trickling into its base and a tiny island just offshore.

The nearest house was over a mile away, a good distance considering the otter's fractious relationship with man.

The next day I packed for my trip, with the very longest of my zoom lenses and plenty of waterproof gear, finally adding a thermos and some sandwiches as I was expecting a long wait. I waddled to the Land Rover swathed in fleeces and Gore-tex. Although it was only early evening, already there was real bite in the wind that whistled off the channel, and dark clouds were massing over the hills above the bay over Applecross. Reuben was duly locked in the bothy, as the one thing you really don't want when out looking for otters is a gigantic black dog helping you out. Otters have very good reasons indeed to fear dogs, having been hunted by packs of otter hounds almost to the point of extinction. I expected to be back within a couple of hours, and had fed him a large plate of food which he was now digesting, snoring and snuffling as he lay by my bed.

It was a short drive along the coast, the edge of the land marked by white foam and spray, with the cliffs, rocks and bays dotted with bracken and heather, rustling in the wind. The island of Rona looked impossibly dramatic in the distance, with dark clouds glowering over the rugged spine of her mountains, occasionally split by shafts of bright sunlight that danced over green fields before being extinguished by the rolling mass of gun-metal moisture overhead.

The bay was actually two separate coves, split by a dramatic rocky headland that divided the onrushing waves, breaking them in two with a percussive explosion of spray. The smaller cove consisted of a white sand beach, the waters of the shallows a startling blue, akin to Bombay Sapphire gin.

The larger cove was more dramatic still, with broken boulders scattered on the shore, a testament to the power of the storms that raged against the buttress of the coastline. Creating the far wall of the bay was a small cliff, angling gently into the water from a high point on the land, finally plunging beneath the seething surface a hundred metres offshore.

I made my way to the central headland and was immediately surrounded by the unmistakable traces of otters. At my feet were lacerated crab shells, broken sea urchins, and fish bones fallen into crevices in the rock. A flat area at the top of the rock created a natural table, and sitting in its centre was an otter spraint, waste matter that marks an otter's territory.

In the world of the otter, a spraint is far more than a bowel movement. It is a love letter, a keep out sign, an email and a postcard wrapped into one. Contained in the spraint is all manner of information on the age, breeding condition, diet and sex of the otter. I had read that they smelled very much like jasmine, and so – feeling somewhat self-conscious – I knelt on all fours and sniffed cautiously at the dark mass before me. It smelled, rather disappointingly, of dung with a touch of mashed crab in it.

Undaunted, I found a suitable gulley out of the wind that overlooked what was clearly a well-utilised dining table, unpacked my gear, dropped my chin into my chest, stuck my hands deep in my pockets, and prepared for a long wait. I glanced up briefly to ensure my camera was pointing in the right direction, and there, looking at me with something approaching outrage, was a large male otter.

He was in the water not ten metres off the edge of the flat

rock, so perhaps twenty metres from where I was sitting. He was holding a rather miserable-looking fish sideways in his jaws, and I suppose his expression was similar to what mine would have been if I had emerged from my kitchen at home in Bristol carrying a large steaming plate of food, only to find someone sitting on my dining table pointing a camera at me.

He lifted his head higher out of the water, and coursed from side to side, gliding without effort in one direction, then turning on his own length to pass back the same way. The swell this close to the shore was substantial, lifting and dropping him into and out of view, presenting a comical series of snapshots of beady cyes and wet whiskers. Behind his broad, flat head was the occasional glimpse of a glossy brown back, with a long tail acting as a rudder from which a V-shaped wake trailed.

He finally rolled onto his back and began to devour the fish, his tiny, dextrous paws juggling and tugging at the slimy body whilst his jaws clamped on the head. His creamy belly became a temporary table top, with pieces of the fish placed periodically on his chest as others were disposed of. Very quickly the fish was divided into its constituent parts and consumed.

Dinner over, he gave one final glance in my direction, rolling over and sculling along the surface before slipping elegantly away to recommence the hunt, leaving a trail of bubbles. These I craned my neck to follow until they too vanished, leaving only the crackling wave tops, the geysers of the spray and the plaintive cry of a nearby gull.

Of course, I stayed where I was for another hour, particularly as I hadn't taken a single frame of the otter as he had rolled and fed before me. The cold seeped into my bones, and the wind sought out any gaps in my clothing, stealing the warmth within, carrying it gleefully away over the rocks and heather. I was alone with my thoughts and the lingering memory of my first good look at a wild otter feeding in the open sea, a mental image I returned to again and again as the shadows lengthened around me.

I finally accepted that the otter would not reappear, doubtless moving his hunting activity for the evening to a place far from the presence of man. I stood and gathered my gear together, grimacing as cold joints creaked and stiff muscles twitched, and began the walk back to the car.

Gavin Maxwell, the man whose writing on Camusfearna first introduced me to the romance of the west coast, was bewitched by the otter – at the heart of *Ring of Bright Water* were his otters, Mij and Edal, the epitome of the marriage of the land and the sea. The otter seems to me to be the wildest of all animals, effortlessly harnessing the power of the ocean, a piece of flotsam riding the currents and eddies, looping, rolling and spinning in abandon, somehow transcending the demands of the daily struggle for survival to treat the sea as a playground.

Perhaps Maxwell found in his otters a creature with which he connected in the most profound way. This was a creature he tamed and yet never truly subdued, something with a veneer of civility that remained absolutely wild, an animal persecuted and hunted that had defiantly withstood all attempts to eradicate it. In the process, it remained resolutely exuberant and free – something that eluded Maxwell throughout his life, no matter how remote his travels became. This meeting of a troubled man and persecuted animal spawned one of the finest natural history books ever written and, as I walked away from the wild bay that is the otter's home, for the first time I felt that perhaps I understood why.

5 July

The Glamaig Hill Race

Although in my time in the croft I was picking up a certain amount of advice on the phone, there was nothing quite like a face-to-face discussion to really sort out the more complex issues. As such, I would frequently find myself climbing into the Land Rover for the long drive to Keith and Rachael's croft on Skye. Their patience seemed to be limitless, as did their supply of whisky and, once Keith had talked me through the issue of the moment, we would open a bottle and the stories would begin.

Keith had spent his entire life working the land, and had developed an affinity with the terrain and wildlife that could only come from decades of experience. He also had a mischievous streak and would tell tales of the absurd scrapes he had engineered for himself as a wild youth, imbued with a well-developed sense of adventure, and a poorly developed sense of self preservation. In turn, I would tell my own stories of travels round the world, of massive shadows in the ocean and explorations of distant shores.

Whilst returning from one such trip, I called in to the Sligachan Hotel at the foot of a great, glowering cone of rock called Glamaig. Stepping out of the car, I craned my neck upwards to the peak, shrouded in mist, atop steep slopes scarred by huge fields of scree. The shoulders of the mountain were a mixture of craggy rock faces and patches of grass, clinging to the edge of slopes that plunged into the swampy ground at the base. It was a beautiful, wild, imposing scene, a monument to the wilderness of Skye that dwarfed any of the puny efforts raised by man in its vast conical shadow, which made me feel simultaneously insignificant and exhilarated.

A waitress brought me a plate of stew in the bar, and I commented on how spectacular the mountain was.

'Oh yes, it's absolutely amazing, isn't it? I've worked here for many years, and I never tire of that first look when I turn up for work. The race is certainly not for the fainthearted.'

I looked blank, a spoon brimming with stew hovering halfway to my mouth.

'That's right,' she said, 'every year folk come from all over the country to race up it and back down again. It's a great day, although I'm not sure I'd be up for that sort of thing myself.'

I lowered the brimming spoon back into the bowl.

'Now,' I said, 'I want to be completely clear about this. People come – voluntarily – from miles away to run to the top of that monster out there beyond the car park. The prize is . . . ?'

'Oh, we give the winner a bottle of whisky, and sometimes second and third place get a woolly hat.'

'A woolly hat?'

'Aye. Although we only do that on certain years.'

This plainly demanded further investigation, and after my meal I sought out the hotel manager who told me the history of the race.

The race was first run, or so local legend had it, in 1899 by a Gurkha named Harkabir Thapa. He had been climbing in the Alps with the famous adventurer General Bruce in the company of Norman Collie from Skye. Collie had invited the General and Harkabir back to Skye, and the sight of Glamaig had proved to be irresistible. Harkabir's first attempt was carried out in bare feet, with the climb and descent completed in an hour and a quarter. The local estate owner – the splendidly named McLeod of McLeod – refused to believe that anyone could conquer the mountain in this time, and argued with the local ghillies who had witnessed the event. To prove the point, General Bruce sent Gurkha Harkabir back up the mountain (Gurkha Harkabir's comments or thoughts on this particular development are not recorded), this time in shoes. Having knocked off twenty-five minutes from his previous time, Gurkha Harkabir was then presumably allowed to go off for a bit of a lie down, and the legend was established.

It wasn't until 1987 that David Shepherd, a member of a local running club, had the idea of restaging the event. He joined forces with the Campbell family, the new owners of the Sligachan Hotel, and their combined efforts saw a small number of runners agree to take part. Tragically, David was killed in a car crash only days before the race, but his brother still competes ever year in his memory.

Having paced the car park for a few minutes, peering up

at the mountain and muttering darkly, I decided it would be bad manners not to give it a go. I resolved to train like a madman and return in the silkiest of shorts, lean and mean with thighs like vacuum-packed pythons, and bound up Glamaig with the rest of them.

Sadly over the next few weeks my training didn't proceed quite as planned, and seemed in the main to consist of eating a great many pies, sitting in the extension drinking tea, and trotting up the dune once a week.

The day of the race duly ambushed me, and with a sense of disbelief I found myself climbing out of the Land Rover into a car park full of lean, pinched people doing complicated warm-up exercises. The majority of them looked like something found in a peat bog by excited archaeologists, the sort of hollow-cheeked cadaver that made the national press terribly excited because they were wearing the oldest flip-flops ever discovered. Most of their upper bodies consisted of lungs, and most of their lower bodies consisted of thighs. The vests they wore hung on the racks of their ribs, whilst the shorts were invariably split to the hip and were of the silkiest shimmering fabric known to man. Only the shoes looked like they belonged in the mountains – much like light versions of walking shoes, wrapping around the base of the ankles and with a heavily tracked piece of tread curling up over the toes, causing me to glance down in alarm at my feather-light, air-soled, snowy-white trainers. My only consolation was the sight of a rather fat man wearing a kilt in the far corner of the car park, self-consciously leaning over to touch his toes then standing to catch my eye, a glance of mutual sympathy and understanding passing between us.

Amongst the locals, Glamaig is described as a hill, and indeed in all the race publicity 'The Glamaig Hill Race' is its official title. It's not a hill, it's a mountain. I'm not quite sure what the definition of a mountain is, although I'll wager there is some cartographical definition of something having to be a certain height. Well, I don't care. I know a mountain when I see one, and the massive, pointy, mountain-shaped thing in front of me looked pretty damn close.

Although they seemed to be from another species entirely, the other runners were friendly enough, spotting me as the rookie I was and coming up to offer advice and support. The general feel was of welcoming another maniac into the fold, with sympathy and just a trace of smugness in evidence from a shared knowledge of what awaited me as soon as the starting hooter sounded. The best piece of advice was 'don't go too fast on the uphill bit', as apparently one needed to save energy – and coordination – for the downhill leg. Glancing up at the mountain, the prospect of my race being ruined by me going too fast on the uphill bits seemed moderately remote. There was every chance of my race being ended by helicopter evacuation or premature death, but not by going too fast at any stage whatsoever.

Soon I was lining up on wobbly legs, looking at the starter in the forlorn hope that he would announce that the race had been called off because someone had realised that the entire thing was ridiculous. He simply smiled the widest of smiles and held up an air horn, and with a loud blast the most painful hour and a half of my life began.

The initial stages were actually rather pleasant, a gentle bimble along the road leading from the hotel, and several easy undulations leading to the foot of the mountain itself. I even started to enjoy myself, overtaking a couple of people on the moderate slopes, unaware that this was all part of a fiendish plot to gently sap my energy before the physiological Armageddon of the main slope. The same people I passed would soon stride past me, a gibbering, heaving, hacking, stumbling ruin of a man.

It was only as the main slope began that my body kicked into top gear, the needles began to flicker into the red, and the pistons started to really thunder. After twenty minutes of excruciatingly hard work, I paused to rest my elbows on burning thighs and heave in several lungfuls of air, the blood roaring in my ears. Glancing down, I saw a line of runners below me, all bent over, and all climbing inexorably in my direction. Some were on all fours, hands gripping the slope as their feet dug into the grass. Even as I turned back to the hill, so steep that the grass was only inches from my gurning features, they were reeling me in. Glancing upwards was even more soul destroying, as I saw a huge line of people pushing determinedly for the summit. This would not do at all, and I began zigzagging furiously to try to take the steepness out of the more savage sections of the slope. Sweat was now streaming from every pore, my heart pounding fit to burst, and my progress was marked only by the sawing, hacking sound of my breathing.

'Keep going, nearly there!' a nauseatingly chirpy voice trilled, the source being a bright-eyed runner of indefinable age who was climbing steadily past me.

'Gtweesh hoomph haaarkgh,' I said by way of reply.

I was approximately a third of the way up.

I won't dwell on the remainder of my skittering, clawed, undignified ascent. Suffice it to say that when I reached the summit I was a profoundly different man from the one who had set out an hour before – somewhat wiser, rather thinner, a lot redder, and infinitely humbler.

And so began the descent. I would liken this to downhill skiing on very steep loose rocks after someone has been vigorously hitting your thighs with a piece of timber for the last hour. Further excitement was added by the fact that the sweat was still streaming into my eyes, giving me a vague blurred impression of the jagged terrain rushing towards me. Catching a toe end on a rock at this stage is ill-advised. Although this would lead to an excellent time, it would mean doing the remaining several hundred metres on your face, pulled along by gravity and pushed along by the gravel-rashed remains of your flailing body.

I ended up following a lady runner for the latter stages. Essentially, wherever she was going, so was I. She might have been a local for all I know, out for a quiet run in the hills. By this stage we were in the small valley between the mountain and the hotel, and as my brain had been boiled in the bag of my skull, I actually had no idea where the hotel was. If she had run home I simply would have followed her, to stand staring glassily into space in her garden until moved on by the police.

I finally stumbled over the line in a time of one hour and twenty-three minutes. Should my life ever need to be justified or judged, I shall point directly to this one hour and twenty-three minutes as undeniable proof that I was made of the right stuff. The fact that I didn't sit on a rock for a good cry halfway up the hill will always be a matter of some pride.

To his lingering credit, the fat man in the kilt also

finished, staggering over the line long after the awards ceremony had finished, with the traumatised look of a man who has stared into the dark places of his own soul. He wasn't last – this honour went to a ruin of a chap in a bloodstained T-shirt who had lost the battle against the combined forces of pointy rocks, gravity and coordination. He walked wordlessly into the bar looking as though he had been vigorously sandpapered.

As I drove away, lifting my leg with one hand every time I had to depress the clutch, I reflected on the peculiar tribe that are the hill-runners. This is a group of people who ceaselessly tour the country, a strange group of latter-day nomads. Their drug is pure pain, their medium is the agony of the steep hill, and the place they seek is the very edge of physiological performance. Despite the horrendous experience of the race, there was something contrary that lurked deep within me that felt briefly satiated. Perhaps I had awoken something that lurks in us all, perhaps I was still delirious, perhaps I had been seriously mentally ill all my life and hadn't known it. Where the sensation springs from is irrelevant – as the mountain faded into the distance in my rear-view mirror, I had a very, very nasty feeling that I would be back the following year.

9

Settling in

As I became more accustomed to the routines of the bothy, so time slipped by and the summer progressed. The azure skies and lazy heat of late spring were replaced with what seemed a relentless series of storm fronts, led by ponderous dark clouds heavy with rain, scudding and grumbling their way down the channel. As the weather changed, so the landscape and animals within altered subtly around me. The stag and his hinds became more elusive, spending protracted periods away from the beach, reappearing only rarely, the stag now shorn of his magnificent antlers. As midsummer had approached, the massive spread of his twelve points had been replaced with a smaller set of velvet antlers, robbing him of his nobility and arrogance, making him look somehow reduced and hesitant. To add to the effect, his hinds had by now abandoned him, moving in a small herd that I occasionally glimpsed on the hill above the dune. This was a time when the stag would refuel, growing new antlers and building bulk and muscle. In mid-August, as the days drew in and the first hint of winter was carried in on the wind, so an ageless clock within would decide that it was time for battle, flooding his body with testosterone and nipping off the blood supply to the velvet covering of the new antlers, which would then be shed to reveal the wickedly sharp weapons of war beneath.

The shallows off the beach were now a cauldron of life, with every trip out on the boat resulting in a glittering haul of mackerel, their firm bodies drumming in their death throes in the fish box. Gulls and oystercatchers worked the shallows, stalking the waterline with delicate steps, periodically probing the sand or squabbling over some piece of jetsam rolling in the waves. The clear, calm water next to the rocks of the headland were full of tiny, darting shoals of juvenile fish, cowering close to the shallow sea floor, only a tiny fraction of whom would see out their first summer.

The vegetation on the hill behind the bothy ran amok down the slope that led to the road, a lush waterfall of bracken jostling for space with the heather, whilst tall foxgloves nodded their elegant heads above the mêlée. The true Scottish heather finally began to bloom, a mass of pink flowers coating the hillside in garish mats. The bog myrtle behind the rocks at the far side of the bay began to spread its leaves, long regarded as one of the finest repellents for the midge. I would wander over to the plant and crush the leaves between my palms, smelling the sweet, oily scent. (This has also been a traditional way of clearing the head, something I tested after a long night in the pub with – sadly – very little effect.) Wild strawberries came into fruit at the edge of the dune, producing tiny, sweet versions of their commercial cousins. Wandering with Reuben along the beach in the morning, I would dart into the vegetation to pick them, gorging myself before returning to the extension for a cup of tea, hands sticky with juice.

On top of the dune, the rowan tree, that most mystical of all trees in Scottish folklore, became gravid with bright red berries. A rowan was planted behind every crofter's cottage, and was considered to be the guardian of the house, never to be pruned or felled under any circumstances. Maxwell himself was convinced that the run of bad luck that saw him driven out of Camusfcarna stemmed from a moment when he was cursed beneath a rowan. The unlikely source was Kathleen Raine, his companion and friend, whose 1952 poem 'The Marriage of Psyche' included the immortal words 'He has married me with a ring, a ring of bright water / whose ripples travel from the heart of the sea' – a phrase that would resonate around the world as the title of Maxwell's best-known book. Their relationship, though platonic, was passionate and fired by their mutual creativity. Banished from the house after a particularly tempestuous fight in 1956, Raine made her way through the storm that raged outside, laying the palms of her hands on the twisted trunk of a rowan that groaned and thrashed in the wind. As the thunder growled and rumbled and the rain hissed on the shore, she lifted eyes wet with tears to the dark skies above, cursing Maxwell by shouting, 'Let Gavin suffer in this place as I am suffering now.' Within the next few years his pet otter was killed by a workman, his house was destroyed by fire, and he was diagnosed with terminal cancer. Every time I climbed the dune, gasping and sweating on reaching the summit, I would pat the trunk of the tree respectfully, and was fastidious in never allowing Reuben to raise his leg against the whorls of its gnarled stem.

Four months before, I had been genuinely nervous about taking on the bothy, feeling somewhat daunted at the prospect of the challenges ahead. My life had always consisted of a great deal of travel, to the detriment of relationships, stability, finances and career. The essence of exploration is to seek the new – to find novel and stimulating experiences, places and people. It had become addictive to me, the thrill of the unique sensation, the quest for the unknown. With such a transient lifestyle came a sort of twisted security. I had always had the comfort of knowing that I was a passing figure, moving on to the next encounter whenever boredom or stability reared their ugly head. I had always maintained a get-out clause, an exit door through which I would slip effortlessly into the next phase of my life. So arrival at the bothy was a commitment, not only in terms of time but also in terms of challenging the very essence of who I thought I was, and indeed who I wanted to be.

We are all perpetually bombarded with carefully crafted messages from advertising campaigns, a white noise assuring us that if we purchase a certain item, move to a certain type of house, or adopt a certain lifestyle, then happiness and contentment will be ours. When it doesn't work, our quest continues – a never-ending mission to secure the latest car, mobile phone, iPod or flat-screen TV. A major part of my sojourn in Scotland was to answer a long-held suspicion that the quest for happiness had a simpler solution – that life really boiled down to being where you wanted to be, discovering who you really were, and connecting in the most fundamental way with the environment around you.

I had begun to suspect that who I really was – or at least who I really wanted to be – was the practical outdoorsman, someone who could whip out a multi-tool to tackle any problem, build a fire in moments, problem-solve using the most basic materials, or effortlessly bring a boat alongside a quay in a brisk squall. On previous trips and expeditions, I had stood on the shoulders of giants, relying on the expertise and practical ability of others. At Sand Bay, from the moment of arrival when I had picked up the hammer and chisel to strip off the old roof, I had been forced by circumstance to become a practical man, a hands-on jack of all trades pushed by necessity to tackle day-to-day tasks that had – up to that point in my life – seemed utterly beyond me.

There are undeniably two faces to the west coast of Scotland. The first is the babbling burn, the soaring eagle, the stag in the mist – enduring images that feed the soul, and are the classic impression of the Highlands. The other side is darker, when the clouds roll in and the windows rattle in the panes for day after day. In this respect, Maxwell's choice of the remote reaches of the west coast was, perhaps, a strange option for a man who seemed to be dogged by misfortune and intense introspection in the latter stages of his life. This wild shore, with short days, rolling clouds and bleak seas beating on harsh rocks, did not seem to me to be a good place for anyone who may be mentally frail. It is all very well seeking yourself, but not when you find that ultimately you don't enjoy the company you may end up keeping. I had felt this keenly as I journeyed north, aware of the gamble I was taking in seeking my own company so desperately, knowing that I would find the very essence of self – a distinctly risky business for a man in comfortable middle age.

Much as I enjoyed the early stages of setting up the bothy, there was an underlying feeling of discontent, a lack of fulfilment that I couldn't quite fathom. Perhaps this was because upon my arrival I had rushed the process of acquiring knowledge of the environment around me, with my head in books, or my eyes to the sky, or my backside in the air and my nose to the ground. Physically, I became drawn and lean, burning fuel perpetually as I trundled around my daily tasks, internal engine gently humming. By the end of the first month, I had reached the last hole on my belt, boring another in the low light of an evening crouched in the shell of my new home.

It was only as time progressed that I realised the process was more gentle, akin to osmosis, as an awareness began to arise of the changing face of the ecosystem around me. A true understanding of the landscape seeps slowly through the soles of your shoes as you walk the hills and rocky shores. Thus the true crofters and fishermen of Applecross had a level of knowledge I could never hope to acquire – an instinctive understanding that was burnt on their hard drive, so profound that even they could not know its depth and scale. It is only by becoming part of the system, by sitting quietly and observing, by walking along the edge of the shore, that I grew to really know the plants and animals around the bothy.

Then, and only then, did I begin to feel a rising contentment and certainty that I was where I wanted to be, a sense of arrival at the end of a very long journey. It is said that we are all instinctively drawn to our natural environment, the place of our original diet, our evolutionary home. As I walked the wild shores of the west coast, I felt a stilling of the restless traveller within, replaced with a sense of homecoming I had simply never experienced elsewhere on my travels. Turning my hand to the more practical aspects of life on the croft had also proved to be a joyous experience. Throughout my life I had measured myself against standards drawn up by others – school, the Royal Marines, university, expeditions, television – and yet here I had perhaps found the true measure of self, when the only asset is the humility to seek local advice, the main tools are the brain and your own bare hands, and questions are asked every hour of every day. In answering those questions, and attempting to understand the gloriously complex environment around me, I had at last found a contentment that had always remained tantalisingly elusive. The irony was not lost on me – after a lifetime of travels and travails, the most profound and satisfying adventure I had ever undertaken had meant staying in the same place.

*

Just as I learned much about myself during my time at the bothy, developing skills and exploring my capabilities, so Reuben was also on a journey of his own. Watching him grow in confidence and strength was one of my greatest pleasures. Since arriving at the bothy, he had grown into a very big dog indeed, weighing about half as much as I did, with a massive dark head, glossy flanks, a broad powerful chest, and fearsome jaws that I had seen splinter wood and bone in seconds. He became a magnificent, happy animal, having thrown off the final memories of the rescue dog he once was.

Reuben's overriding characteristic was his utterly indestructible good nature. He was simply the most gentle, genial animal I had ever encountered. He was something of a celebrity amongst the livestock. I would often find him sitting at the gate to the pig enclosure, with Gemma nuzzling him through the bars, the two of them chatting away in a series of grunts and snuffles. The chickens had lost their fear of him, and he would sit at the fence as they pecked at the ground only metres away, his head on one side as he watched them with what seemed to me total absorption. Even the sheep seemed to tolerate Reuben, lying contentedly alongside the deer fence as he trotted along its length.

He did have one rather bad habit, though. When overexcited – a scenario usually involving a stick, some sand, possibly some sea, and lots of shouting and waving – he would leap up to nip the holder of the stick. This is classic behaviour in puppies playing with their litter mates, and is at best an irritant when it comes to young chinchilla and miniature poodles. It's a slightly different matter when the nipper has jaws you could wrap round a beach ball, and a number of locals – all of whom loved Reuben – soon had a small tear in their left sleeve, something that became almost a badge of honour around the village. This characteristic passed with age, but for the first few months of our time together in the bothy throwing a stick for Reuben was a very exhilarating pastime indeed.

Reuben's day would begin with a shake of the head, accompanied by a machine-gun rattle of floppy ears that

ABOVE The Jackson family — my mentors and friends.

RIGHT Tristan Jackson narrowly wins the gurning competition while out fishing.

BELOW A well-dressed Kayak Mike looks smug as bad weather sets in.

All creatures great and small on my doorstep, in the deep blue channel that stretched across to Raasay and Rona — a constant source of wonder and delight.

Sand Bay. One man and his dog shared green hills, golden sand and blue water – paradise for both myself and Reuben.

ABOVE Dave Abrahams keeps a careful eye on the land around him.

ABOVE Into the wild.

BELOW The Royal stag paying a regal visit to the bothy.

Community is what Applecross is all about — never more apparent than on the day of the Highland Games.

resolutely refused to stand up in the approved Alsatian manner. He would then stretch, yawn cavernously, and make his way to the door. First stop – after the morning pee – was the sea, and he would trot down the path to stand immersed, only his head showing. This stage could last a very long time, and on occasion I would finish all my chores to return to the extension and find him still there, shivering with cold and holding a frond of seaweed in his mouth in the vain hope that I would throw it for him.

He loved the water, more than any other dog I had ever encountered. On occasion, when I was writing in the extension and Reuben was yawning and sighing at my feet, he would stand up and amble to the water's edge. With a single glance back at me, he would walk into the sea and swim for twenty minutes or so, often a considerable distance into deeper water before returning to shore, pausing only to give a colossal shake of his body, so violent that his feet would skitter and dance on the smooth pebbles. He would then return to the extension and fall at my feet, snorting and snuffling into sleep.

He particularly enjoyed a wrestle on the beach, and would crouch in anticipation as I approached on all fours, timing my final spring towards him for maximum impact. During the course of these distinctly one-sided contests, he would roll, spin and cavort on the sand, mouthing any portion of my anatomy I was unwise enough to leave unguarded. These bouts could go on for some time, with both of us retiring battered and bruised to the extension – with me reflecting on the tiny glimpse I had just received of his potential power, and Reuben with that bushy tail held aloft like a victory banner.

I had begun to suspect that Reuben had a certain measure of Newfoundland in his lineage, a quality most apparent when we were in the boat. He loved these trips, paws on the tubes of the RIB as we sped through channels and past sheer cliffs. He was calm and measured in the boat, moving carefully around diving kit and fishing tackle, every inch the competent crewman. It was only when one of us entered the water that all hell would break loose.

Before I had an inkling of his seafaring lineage, I took him out on the boat on a sun-gilded afternoon. The sea was glassy calm, so I anchored in a small cove and prepared for a brief snorkel in the shallows. Moments later, clad in a wetsuit and simultaneously juggling mask, fins and camera, I gave Reuben my most stern look, and barked a command to stay before slipping over the side of the boat and finning idly to shore. As ever on entering the water, I felt a massive calm descend on me, the gentle swaying fields of kelp beneath, the flitting schools of sand eels in mid-water, and the sun dancing on iridescent patches of white sand.

I was therefore taken somewhat unawares when a gigantic dog landed squarely between my shoulders. Reuben's most basic instincts had kicked in – this was plainly a rescue situation, and the person in the water was going to be rescued even if he didn't want to be.

Snorkelling and taking underwater photographs is complicated enough: it's rather kit-intensive, with delicate cameras in waterproof housings, strobes, metal supports, wires and curved glass ports. Add to the mix a massive dog furiously paddling alongside throughout, grabbing any available extremity in order to drag you back to the boat, and it becomes almost impossible. After half an hour of running battles, I finally climbed exhausted back into the boat, with Reuben hopping in alongside me having shepherded me in for the last few metres. He gave me a final cursory inspection and trotted to the bow where he curled up and went to sleep – although one eye would periodically open to check that I didn't need rescuing again. From that point on Reuben would remain in the Land Rover when we were diving from the boat.

This proprietorial side to his nature meant that he soon came to regard the beach as his personal patch, not in a protective way but more as a benign landowner. Such was his affinity with the beach that he would lie out in front of the bothy even in the pouring rain, always facing the sand, eyeing the surf and the dunes. His coat would absorb the downpour, the outer layer becoming spiky and matted, whilst the soft inner layer remained bone dry. He was never more contented than when I was working around the croft and he was nearby gnawing on a piece of driftwood, pausing only to lift his head and study the beach, the sea, and the hills beyond. Should anyone appear through the dunes, Reuben would prick one ear up (the other could only rise in the face of a stiff wind) and walk swiftly down to greet them. When there were several people on the beach, he would become quite rattled, darting from one group to another, tail wagging – a conscientious host checking that everyone was having a good time.

Most visitors to the beach were locals, and would look forward to seeing Reuben trotting towards them, greeting him affectionately and playing with him patiently, sometimes for hours on end. There was, however, the occasional group of visitors who greeted the sight of a grinning, jet-black

Alsatian the size of a Shetland pony bearing down on their family group with considerable alarm. This was entirely understandable, and I would try to keep Reuben tethered to the front of the bothy whenever there were kids on the beach that I didn't know. There were occasions when I was slightly too late, and would emerge from the courtyard to see a family speed-walking back to their car, carrying absurd amounts of beach paraphernalia, ushering their kids on ahead, with the dog following delightedly close behind. I saw one father even hurl a large rock at him, and was close enough to see his look of complete bewilderment as Reuben bounded after it, picked it up and dropped it carefully back at his feet.

Reuben's territory extended well beyond the surf line, and he would make regular sojourns out into deeper water on his morning patrols. One such morning, I was walking the shore as Reuben paddled several metres out, effortlessly keeping pace with me, legs whirring and tail deployed as a dark rudder. Without thinking I absent-mindedly picked up a pebble and lobbed it several metres in front of him. The telegraph went from idle speed to flat out, with his sudden acceleration signalled by an increase in his already substantial bow wave and his ears flattening (presumably for maximum aerodynamic efficiency, rather like a jet sweeping back its wings).

He was at the spot where the pebble landed within seconds, and – as I watched with a rising sense of disbelief – circled several times on the surface before deciding that it was obviously on the sea floor. This was plainly unacceptable so – taking a breath and shutting his eyes – he duck dived, his buoyant furry backside bobbing on the surface, back legs pumping in mid-air like the spinning props of a nose-diving ocean liner. After several seconds, he surfaced triumphantly, pebble held in his mouth, spluttering due to the ingestion of several pints of seawater. He swam ashore and deposited the pebble at my feet, coughing and blinking. I was delighted at this development, and the unfortunate Reuben was required to perform this trick on several future occasions to the polite applause of visitors to the bothy. I would stand and beam like a proud parent on sports day, blissfully unaware that everyone – Reuben included – was probably humouring me.

As time progressed, Reuben seemed to become one with the landscape, a magnificent, proud animal that fitted perfectly with the big skies and craggy cliffs that made up his new world. This was never more apparent than when we went walking in the hills, where he would bound on ahead, seeking out the high ground to check for prey, in his mind's eye leading the pack on the hunt. He would periodically circle back to me, urging me on, before loping again into the middle distance. He appeared to blend perfectly with the heather and the bracken, a peculiar genetic memory stirring in me of the days when wolves roamed the hills and forests of Scotland. The red deer we encountered shared the same feelings without the romantic connotations, and they would flee the moment they saw him. This was not the gentle elegant trot of the mildly disturbed deer, but headlong flight, a twisting and turning gallop to a safe distance, before gathering together in small groups to observe the low, slinking carnivore that had appeared from their deepest memories and nightmares. Reuben would always glance back at me when he sighted a deer, with my resultant low growl fixing him in place. He would then circle my legs, peering at the running deer with only the occasional whine indicating the turmoil within.

Reuben was my attentive, unselfish, faithful and amusing companion during my sojourn on the west coast – I could not have hoped for more from any human. At the end of one particularly bleak day, when the rain had drummed and hissed throughout, and the cold had seeped into my bones as I tramped around the bothy, I had finally collapsed on the battered leather sofa in the extension. Overwhelmed by fatigue, chilled, demoralised and yearning for my warm home down south, I turned my back to the seething maelstrom of the channel outside the window, and tucked my chin into my chest, desperate for sleep and temporary escape.

It was only moments before I felt massive paw rest on my shoulder, then a damp nose nuzzle the top of my head. I shifted deeper into the sofa, in no mood to play with dogs. Reuben whined gently in my ear, then lifted his body onto the edge of the cushions, sensing my exhaustion and despair. Within moments he was spread full length, pressed against my back, warming and comforting, my great friend keeping the demons at bay. The two of us swiftly passed into deep sleep as the wind rattled the windows in their frames, an oasis of warmth in the eye of the storm.

There will come a day when Reuben will pass away, as we all must do. When this day comes – and what a bleak time that will be – there is only one place where I will scatter his ashes. His final resting place must be the surf's edge at Sand Bay, the place where he threw off the shackles of his past, and became a strong, healthy, noble dog, his spirit forever preserved in the endless movement of the waves, the cool mist of the spray, and the gallop of the deer as they spring through the heather on the hillside.

*

As July gave way to August, previously laid plans in the croft began to come to fruition. The raised vegetable beds were now virtually out of control, the crops spilling out over the edges of the concrete walls. The hens were now producing at least six eggs a day, leading me to tour the village handing them out to friends. My repertoire of omelettes increased dramatically, including some experimental mixtures that achieved only limited success – if I learned nothing else from my time in Scotland, I can at least say with absolute confidence that a raisin omelette will never catch on.

With the summer holidays now in full swing, the beach was also receiving its fair share of visitors. This meant that Reuben virtually had a nervous breakdown, darting from group to group, tongue lolling and eyes bright, a host with slightly too many guests at the party. Children played in the shallows only yards from the bothy door, as their parents set up what appeared to be substantial base camps on the beach. It is a regrettable characteristic of British holidaymakers that they seem incapable of visiting the beach without windbreaks, mini-tents, towels the size of squash courts, and an array of collapsible furniture.

I was treated one evening to the sight of a very large, rather obese gentleman moving in stately fashion towards the sea. He eventually started to immerse himself, the water creeping up marbled thighs and mottled flanks as he ploughed into the small surf. Eventually, he reached a sufficient depth for his great weight to be supported by the water, falling forward like a tanker down a slip into the Clyde, a substantial bow wave racing away ahead of him. This would have been a mildly disturbing sight under any circumstances, what made it doubly nauseating was that he was stark naked. He eventually emerged to move slowly back up the beach, evolution in reverse as an ocean giant abandoned weightlessness to reclaim the land. Even Reuben glanced up at me in bewilderment as we both stared in horrified fascination from the extension window. The array of extraordinary, graceful and elegant creatures that had passed before the window had been replaced – albeit briefly – by people visiting the beach. It made me yearn for the isolation of the wild days of wind and rain, and reinforced my view that we can, at times, be a singularly unlovely species.

For the vast majority of the time, however, we had Sand Bay to ourselves. Antje, Reuben and I revelled in a return to warm, calm days as the troubled adolescence of summer became a settled maturity, with a return to balmy seas and blue skies.

During one such afternoon, as I tapped away at my laptop with a hot cup of tea beside me, there came a frantic hammering on the extension door. Opening it, I found a distraught woman, trembling with emotion and stumbling over half-finished sentences.

'I can't believe . . . I'm shaking all over . . . feel sick . . .'

The very worst thoughts raced through my mind. A drowned toddler, a savaged dog, a twitching spouse on the sand turning a delicate shade of blue.

I gave her a moment to calm down, and the true tale emerged. It was certainly grim, but in the perspective of the horrors I had imagined, almost a relief.

'I'm really so sorry. I was having a picnic at the back of the field there, when my dog slipped under your gate and . . . ' She paused, her eyes studying the ground once again. 'I think he's killed one of your lambs.'

Sure enough, a quick search of the field revealed a lamb, throat torn open and quite dead, half-hidden in the bracken. By this time, her husband had appeared, a large, florid, friendly-looking man. He wore a flat cap, the classic checked shirt of the country man, and his ruddy complexion meant he had the look of the land about him.

He shook my hand whilst apologising profusely, nonetheless holding my gaze with complete confidence. Plainly used to giving orders, he seemed to be the sort of figure that wore power and influence easily, and I wondered briefly if he was a holidaying executive – a captain of industry escaping the rat race for the sanctuary of the west coast.

I tend to be fairly philosophical about incidents such as these. Admittedly they should have been watching their dog more closely, but then again perhaps I should have been more studious about tending to the gap under the gate. They were plainly very sorry indeed, and were adamant that the dog had never shown the slightest inclination to attack sheep previously. The fact that it was a Jack Russell explained things slightly – a terrier to the tips of its toes, with the instinct to kill burnt deep in its genes. The moment the lamb had bolted at the sight of the approaching dog, an age-old mechanism had clicked into place and the path of the next few moments became irreversible.

The three of us chatted briefly in the extension, before the wife made her excuses and left, closing the door quietly behind her with a final apology. After several more minutes, the husband – who had introduced himself as Mike – also stood to leave.

'I'll get one of my cards from the car,' he said. 'You can give me a call, and we'll talk about compensation for the lamb.'

By this stage he had reached the door and gave the handle a rattle. It remained resolutely shut. He rattled harder, the door merely shaking in its frame as a result.

Mike's wife, still plainly shaken by events, had put the bolt across on the outside of the door. I was locked in a very small extension with a very big man. I glanced out of the window at the deserted beach, and then at the dune behind which his wife had disappeared. We both hammered on the door and tapped on the windows, all the while shouting for assistance. We didn't shout too loud, however, as any form of overt emotion seemed to be letting the side down somewhat. There are many Englishmen – I know I'm one of them and I suspect Mike was as well – who would rather die a slow death than indulge in overt displays of emotion. The Blitz spirit ultimately kicked in; he collapsed onto the sofa whilst I settled in one of the hard-backed chairs, and we began to talk.

Mike told me about the great art of the stalk – his main reason for coming to the Highlands each year. He told me that it was essentially a cull, killing the weak and starving stags on the hill as the winter set in.

'You should see them sometimes, it really is pitiful. I know that stalking may seem purely sport, but it stems from an old tradition that eradicates the weaker animals. It's genetic modification in a way – albeit genetic modification using a powerful rifle,' he added drily.

Earlier I had carried the limp body of the lamb through the gate, dropping it beside the Land Rover for later disposal. With a sudden, horrible moment of realisation it occurred to me that Reuben was outside, off his lead, with the bleeding body still out in the open. I quickly moved to the side window, expecting to see bloodied jowls and scattered entrails. Instead, I was just in time to see Reuben's gigantic form move towards the lamb. He sniffed it cautiously, then took several steps backwards, head tilted to one side. With an exasperated sigh, he sat back on his haunches, glanced around him, then settled onto his backside, one ear in the upright, alert position. He remained this way as I continued to watch, all the while glancing around him, looking back at the lamb, then returning to his vigil.

'He's guarding the lamb,' said a loud voice beside me, startling me out of my reverie. Mike had silently appeared at my side, and was watching the unfolding scene with an experienced eye. 'You've got a good dog there.'

Looking at Reuben, it occurred to me that I was watching one set of instincts – the Alsatian coming to the fore to guard the lamb – being used in response to another – the terrier's reflex to kill.

We returned to our seats, and Mike began to ask me about my time in Scotland. He was particularly interested in my relationship with the locals. He listened with great interest as I explained my experience of the local community, particularly when I talked about the landlords of Applecross, the Wills family. The relationship between them, I explained, and the community itself seemed extremely complex. Warming to my subject, I said that I hoped they realised what a rare and wonderful place they presided over. I became quite animated about the entire subject, before sitting back with a resigned wave of the hand.

'You do wonder,' I said, 'if there is still a whiff of the feudal about the entire set-up. Anyway, I'm rattling on. Would you like a cup of tea?'

Mike's answer was interrupted by the return of his wife, and the unmistakable sound of the bolt sliding back. Mike thanked me politely and stood to leave.

As he was passing through the gate, I realised that I still hadn't taken his number. Rummaging in my pockets, I found a scrap of paper and a tiny stub of pencil.

'Mike, in all the excitement I still haven't taken your number.'

'Ah, of course you're right,' he said, with a slight smile, scribbling down a mobile number. 'And I should give you my full name before I go. It's Mike, as you know, and the last name is Wills. Be seeing you around.'

With an even broader smile, and a last handshake, he was gone.

10

Peaks and Troughs

The mountains beyond the channel glowered at me day after day, menacing or magnificent depending on the weather. They were at their most beautiful as the sun set, changing hue from moment to moment with their soaring slopes lustrous in the light of the dusk, becoming slowly shrouded in the shadows of the night as the sun vanished beneath the horizon. They never truly disappeared though, even on the darkest nights, their solidity and mass supporting the evening sky.

I had never been a particularly keen mountaineer, being drawn more to the echoing depths of the sea than to the exposed shoulders of hills and ridges. At the root of this was a firmly established fear of heights. This had come thoroughly to the fore on a recent television project in Borneo, during which I had been required to be suspended on a gossamer strand of rope over a ninety-metre drop.

When the cameraman was lowered down to join me, instead of receiving the pre-planned pithy piece to camera, he was subjected to the deranged babblings of a madman, wild-eyed and slick with sweat.

Despite these fears, I have always been fascinated by climbing, and read avidly about the great expeditions of the past. I was in thrall to the golden age of mountain exploration, when tweed-clad young men trudged into the mists of the Himalayas, pitifully ill-equipped for the horrors that lay ahead, drawn in by the ice, the rock and the promise of fame. Two British climbers in the 1920s particularly fascinated me: George Mallory and Sandy Irvine, who were last seen 'going strongly for the summit' on Everest in 1924, before being enfolded forever by the clouds, grimly striding into legend.

My view of the mountains actually encompassed a series of ranges, starting in the foreground with the sharp backbone of the island Raasay, leading onto the mountains of Skye, and with the distant horizon dominated by the legendary Cuillin Ridge. The latter was a truly fearsome spectacle, with cols, saddles, spines, ridges and pinnacles jagging through the skyline, their precipitous slopes a patchwork of shadow and scree.

The Cuillin Ridge incorporates sixty-eight separate peaks, and it is the holy grail of the truly fanatical hillwalker. There are places on the ridge where basic climbing skills are required; however, the majority of the journey along the spine of the mountains requires only steely resolve, stout thighs, a basic grasp of roping skills and a healthy disregard for the perils of tiptoeing along a path the width of a mantelpiece, ignoring yawning voids hundreds of metres deep on either side. Personally, I would rather lower my private parts into a jam jar of infuriated wasps, yet several hundred people make the trip every year, and good luck to those that do.

Although the ridge never rises above a thousand metres (which is still a kilometre straight up into the sky, I hasten to add), to traverse its entire length is a very serious challenge indeed. It stretches for a total of thirteen kilometres, has twenty peaks over 900 metres high, and requires that the walker completes a total of 2,700 metres of ascent. The record for this traverse is held by a lunatic called Andy Hyslop, who leapt, clawed, bounded and cackled his way along in three hours, thirty-two minutes and fifteen seconds.

Although I felt the ridge was probably beyond me, I did want to venture into the mountains during my time on the west coast. I wanted to pay my respects to the great serried ranks that faced me every day, to explore both them and me, to tiptoe along the edge of chaos.

Kayak Mike was the obvious man to chat to, and the moment the words were out of my mouth his eyes lit up.

'I wondered when we were going to get you into the hills,' he said, thumping the table for effect so our pints jumped and clinked. 'Give me a date, and I'd be delighted to take you along. I'll even take a day off work. Can't think of a finer way to spend an afternoon than with a terrified posh bloke.'

Mike invariably had a map to hand, and within moments was poring over a series of densely packed contours, mixed in with craggy lines indicating boulder fields and tumbling slopes of scree. He pointed to a tiny dot in the midst of this cartographical bedlam.

'Here we go, Monty,' he said with a huge smile. 'The Inaccessible Pinnacle – perfect for a lightweight such as yourself.' He seemed delighted with the choice, folding the map back into his rucksack with a satisfied chuckle. 'You know what,' he said, warming to his theme, 'we could even use your RIB to access it from the more remote side. Lovely sea trip in, good hard haul to the base of the climb, then a bit of exposure as you scale the final pitch.'

Terms that were unavoidably associated with climbing were now coming thick and fast, too quickly in fact for me to blank out. I had been particularly keen to avoid any plan that involved words such as 'remote', 'haul', 'climb', 'scale' and – worst of all – 'pitch'. Such terms evoked alarming images of crampons skittering on rock-hard ice, pitons being driven into cracks by swarthy Alpinists, lengthy dangles on unravelling hessian ropes, and a slow demise whilst high-altitude choughs peck enthusiastically at one's ice-rimed face.

Several weeks later, I drove the boat on the trailer over the Bealach na Ba, having arranged to meet Mike at the tiny port of Armadale on Skye. Unbeknownst to me, this was to be the start of the very long day indeed. Trouble with both the boat and the trailer meant that what should have taken an hour and a half ended up as a nine-hour epic, with me finally pulling into the car park at Armadale as darkness was drawing in. Mike had already arranged accommodation in a quaint bed and breakfast, and listened to my trials and travails sympathetically before urging me to get an early night to prepare for what he promised was going to be a 'demanding' few hours in the mountains the next day.

Reaching the foot of the ridge would require an hour-long transit by sea the next day, followed by a four-hour walk to the base of the Inaccessible Pinnacle, where the climb would begin. Breakfast consisted of a massive plate of fried food, with slabs of bacon and plump sausages the size of draft excluders. We wobbled our way down to the RIB as it bobbed prettily alongside the floating mooring, threw in the climbing and trekking gear, gunned the engine and, with an explosion of foam and a billow of exhaust fumes, our adventure began.

A particularly glorious hour later, we drew into a small cove framed on three sides by rugged cliffs that hissed with a network of small waterfalls. Our trip over from Armadale had seen us knifing through flat water, without a breath of wind to ruffle the silky surface. Within moments of leaving the harbour we had sighted a basking shark, the two tiny triangles of its dorsal and tail fins sculling gently along parallel to the shore. This is an entirely misleading sign of the size of the animal beneath, which can be the same length and weight as a double-decker bus. Following Mike's shout and outstretched arm, I had swung the RIB round towards the fins, only to see them disappear in a flurry of spray, the water bulging and swirling as the animal sounded. We waited to see if it would reappear, but this was still early in the season for the sharks and, knowing that the really dramatic encounters would take place later in August as their numbers increased, we swung the nose round to continue our journey. The cliffs and bays of the mainland had receded behind us, shimmering in the heat haze, before the more rugged spine of Skye had grown stark on our bow. Pulling the RIB into the cove, the gin-clear water giving us the impression of flight over the white sand and swaying weed, we anchored alongside a small slip. Mike quickly and competently prepared our gear and, with a final glance at the sun, hefted his rucksack onto his back to begin the walk into the valley ahead.

'I can't believe this heat,' he said, wiping his brow as I drew up alongside. 'As a native Scot, I should warn you that if it gets any warmer than this I may well spontaneously combust.'

I was already glowing bright red, blinking sweat out of my eyes as I walked. Mike was setting a very brisk pace indeed, skipping from rock to rock and opening his stride along the flat. He had spent a lifetime in the mountains, and already I could see his shoulders begin to relax, his head lift and a slight smile appear as he walked through one of the most spectacular valleys in Scotland.

Next to us was the great shining mirror of Loch Coruisk, stretching away to the foot of Sgurr Alasdair in the shimmering distance, the air writhing and pulsing with the heat. There was a cathedral-like feel to the scene: a massive amphitheatre of craggy rock, green slopes, waterfalls and tumbled scree beneath a soaring roof of blue sky. The ridge dominated the skyline, with cols, spines, arches and cliffs weaving a line through the dense cloud that was draped on the crests – a result, said Mike, of the air rising sharply up their steep sides to crash into the colder air at altitude.

When we finally reached the base of the slope, Mike suggested we fill our water bottles before the final push towards the ridge. The stream alongside was of such clarity that the fish in the depths of its pools looked somewhat out of place, as though hovering in mid-air over the smooth stones below. The water had a peculiar blue quality, almost iridescent, a mix of pure mountain water and minerals washed from the glacial rocks that had guided it to the valley floor. Plunging my hand into the water, so cold it made my skin tingle, I filled my water bottle as I stared up towards the pinnacle.

The Inaccessible Pinnacle sits in the midst of a saddle on the ridge, looking like a shark fin ploughing along the skyline. The path towards its base required an ascent up a grassy slope, then several boulder fields, then a large area of scree before the ridge itself. Letting my gaze wander up the route, I was struck by the stark nature of the landscape. The Cuillin Ridge and the valley of Loch Coruisk is a result of geological uplift and glaciation. The landscape had the look of a battlefield, all deep scars and shattered rocks that spoke volumes for the vast forces to which it had been subjected. Ice and stone had fought in a series of pitched battles, leaving deep trenches and scattered shrapnel of rock shards, with gigantic boulders that leaned drunkenly in defeat. It seemed to me that, even in this new age of extreme tourism, of Gore-tex and gaiters, that somehow modern man does not belong here. I was viewing a primeval landscape, nature in its most basic form, which appeared to reject and belittle any attempts by man to dominate or subdue it. Every nation needs a wilderness, somewhere its people can visit to escape completely the trappings of civilisation. Such experiences remind us that not so long ago we were creatures of the wild open spaces, superbly equipped to survive and prosper in such a savage landscape. It represents an opportunity to touch something basic within ourselves before we flee back to what only seems essential to our survival in the world we have created.

Mike had already struck out for the base of the first slope, and I hurried to join him as he began to move up the steep gradient. We were both carrying about fifteen kilos in weight, consisting of climbing, filming, photography and basic survival gear. I was also carrying something that I knew weighed precisely two pounds – a massive lump of almost a kilo of mature cheddar (strength four). Sadly, this hefty slab took up all the space I had set aside for food.

I produced this on our first stop, next to a babbling waterfall at the base of the boulder field. Mike looked at me, somewhat startled, as I took a bite out of the cheese, then shook his head as I offered him a chunk.

'Interesting choice of mountain food you have there, Monty,' he said, with a quizzical lift of an eyebrow. 'I suspect it's just what your dehydrated, fatigued muscles are crying out for right now, a huge lump of cheese.'

'Ah yes, the cheddar. You see, they were offering a discount if it was bought in bulk, so I thought I'd get some and bring it along.'

'Good to see its strength four, anyway.'

'Oh yes. I've tried climbing mountains on strength two and you just feel weak and listless.'

Fuelled by cheddar and mountain water, we pushed on up the hill. The going was very hard indeed – with the steep grassy hill soon giving way to boulder fields. The latter I found particularly nerve wracking, a Jenga-style pile of massive rocks, each teetering on another, creating a house of cards that I felt would tumble with one wrong step. Mike moved quickly and cautiously ahead, testing each footfall, and occasionally stopping to peer up at the chaotic jumble ahead of him, seeking out the safe line. He told me between heavy breaths how he had once dislodged a boulder the size of a fridge, darting out of the way to avoid it, only to trip and fall, watching it roll towards him.

'It was precisely like a scene out of a Hollywood film, except that this particular boulder wasn't made of polystyrene. I defied some basic laws of physics that day, springing upwards and moving sideways several feet in a millisecond, all from a lying position.' He shook his head at the memory. 'Still not quite sure how I managed it, to be honest.' He chuckled drily and continued his measured ascent. The boulder field groaned and creaked beneath me, a large man in heavy boots walking on tiptoes.

After several hours, and at least a quarter of the cheddar, we paused at the base of the final scree field that would lead us to the base of the pinnacle itself. Drinking deep from my water bottle, I took in the great expanse of the valley below me. We were now at an altitude of over 600 metres, and the glacial history of the landscape was plain to see. Boulder fields marked the end of the progress of the glacier itself, a moment frozen in geological history as the ice receded for the final time. The stream twisted through the base of the valley, fed by tiny tributaries that seeped out of the steep slopes in glinting rivulets. The ridge itself was now almost completely shrouded in cloud, causing Mike to glance up in concern for the first time.

'We need to be ready to turn back if required,' he said. 'This is one place we really don't want to be if there is the chance of an electrical storm. Standing triumphantly on top of the Inaccessible Pinnacle with your arms raised could result in a very interesting hairstyle indeed.'

We were soon at the base of scree, smaller stones that

tumbled from the ridge itself in a thick mass. Walking upwards through the scree was tremendous work, much like walking through thick snow. Even Mike's pace slowed, with both of us pausing occasionally to lean on our knees, heaving in deep breaths and peering upwards. At last we reached the ridge, which was enfolded in thick mist. As we entered the clouds, the atmosphere became more sinister, with slippery basalt rock underfoot.

'This is a very good opportunity indeed to screw up,' said Mike, a shadowy outline ahead of me. 'Miss your footing here and you'll end up skidding down 900 metres of scree on your backside, which would leave something of a rash.'

His voice sounded strangely disembodied emerging from the mist, somehow tinny and leaden, diminished by the height and enfolding gloom.

I was so focused on placing my feet correctly, with only the occasional wide-eyed glance around me that I bumped into Mike when he finally stopped immediately ahead.

'And this,' he said, with a generous wave of the arm, 'is it!'

I could see absolutely nothing, save for a barely perceptible darkening of the fog bank in front of us. Screwing up my eyes and craning my head forward, forehead furrowed, I could just make out the outline of a hulking monolith of rock.

Mike quickly strapped me into a harness and placed a helmet on my head. I always – always – look completely absurd in any type of helmet. It may be something to do with the unusual curvature of my bone structure. It may be the fact that my head is quite generously proportioned. Whatever the cause may be, helmets tend to balance in an amusing manner on my skull, a considerable distance from my face a long way below, wobbling from side to side when I walk. Mike took several paces back, peered at my helmet with a frown, and then moved forward again to tighten the straps further. The only effect this had was to make my eyes bulge and to briefly cause me to claw at my windpipe. He loosened everything again, shrugged, and turned to lead me to the foot of the Inaccessible Pinnacle.

Mike had explained to me earlier that the name of the pinnacle stemmed from the Victorians' penchant for the overly dramatic. 'It's much the same as the clan Tartans,' he said. 'The invention of a society smitten with the Highlander – although we're all such magnificent fellows that you can hardly blame them.' He added that as a climb, this really was very easy indeed, although he did mention a certain level of 'exposure'.

My heart sank. Exposure is the embodiment of every vertical nightmare for me, encapsulating everything I find unnatural about climbing. The term essentially means a feeling of vulnerability as yawning drops beckon with their cavernous maws. It means sweaty palms slipping on smooth basalt, and the realisation that you are a tiny speck clawing up the face of a giant. Add to that the feeling of being revealed as an utter climbing fraud, and Exposure becomes a particularly apt term.

'Right, Monty, here we are,' said Mike. 'I'm going to go on ahead, get in a good position for the initial belay, and then you can climb up to where I'm sitting. Enjoy this my friend – a rare departure from sea level for you.'

After a final quick check of my harness, and a few last safety tips, Mike turned to the rock face, placed his hands on the rough stone and, within seconds, had vanished into the mist, a wraith ghosting his way upwards into the ether.

This left me alone at the foot of the face, contemplating an uncoiling length of rope, and the first slithering of fear within my belly. My forced inactivity – for the first time that day – gave me pause for thought, and a moment to consider the peculiarly individual nature of fear.

One of my passions is public speaking. I'm aware that this probably speaks volumes for all sorts of hidden personal issues, requiring as it does a certain approval-junkie mentality, basking in the spotlight of a captive audience. Nonetheless, give me an audience of several hundred, a stage to stride, a story to tell and there's no stopping me. The adrenalin surges, the senses sharpen, and I come alive.

Several years ago I gave a talk at the Royal Geographical Society about diving. Accompanying me on the bill was a famous cave-diver. Cave-diving is the most ludicrous activity imaginable, requiring the participant to push deep into the heart of the Earth, penetrating the compressed arteries of unforgiving rock passages, squeezing through constrictions in pitch blackness, sometimes removing kit to push it ahead, supported entirely by a wheezing, hissing, clanking collection of valves, cylinders and tubes. All cave-divers take along a second set of kit as a back-up, but should this fail the diver dies. No ifs, no buts, no 'how about if we try . . . ' – kit failure results in a dark, lonely death as far from the warmth of fellow man as it is possible to be. This fact alone means that cave-divers must be nerveless, a breed apart that voluntarily enter a cold, echoing, unforgiving world that the rest of us can only begin to contemplate. Lose control of yourself, give in to panic, and the result is death. The spectre of your own mortality is your dive buddy, whispering in your ear and constantly threatening to enfold and overwhelm you.

One of the most famous examples of this occurred in 1978, when three cave-divers pushed into a network of passages and tunnels deep beneath the Yorkshire Dales. The team consisted of Oliver Statham, Geoff Yeadon, and German cave-diving expert Jochen Hasenmayer. These men were at the cutting edge of cave exploration, pioneers pushing ever further into the stifling darkness, forcing themselves and their equipment deeper and deeper into the cold, hard, unforgiving heart of the Earth.

As was the custom during this period of cave-diving exploration, they entered the cave one at a time to avoid getting in each other's way in the constricted passages. Their rationale was that a companion would be of no use in a serious incident anyway, as they were too far into the caves for two people to get out on one person's air supply. Horror stories abounded of one diver attacking another as the last breath trickled from their lungs, desperate to maintain life at any cost.

As a cave-diver of near legendary status, Hasenmayer went first, with Statham following later, then finally Yeadon. Yeadon was surprised as he worked his way through the labyrinth of pitch-black submerged tunnels and sumps to come across a returning Statham, who wrote the ominous words '3000. Small with back and sides. No Jochen. Trouble?' on a slate. Loosely translated, this meant that he had achieved a penetration of three thousand feet (900 metres). The latter half of the message informed Yeadon that Statham had not encountered Hasenmayer on the return journey. The final word 'Trouble?' hinted at the implications of this.

Yeadon decided to push on into the cave to try to find their companion.

He eventually saw the glow of Hasenmayer's lights in a sump, a widening of the normally constricted body of the cave. Following a line towards these lights, he then saw a tiny movement. Hasenmayer had become briefly disorientated, and was now stuck on the other side of the main tunnel, with only a small gap in the cave wall no larger than a letterbox through which his hand was fumbling, trying to find life in the void.

Yeadon grasped Hasenmayer's hand, squeezing it to reassure him, a moment of human companionship and warmth in the darkness. It seemed that Hasenmayer was completely doomed, his air running low, lost in a parallel cave that had never been explored, many hundreds of metres from the surface. Although the grip remained firm, Yeadon felt he was holding the hand of a dead man.

After several minutes, Hasenmayer realised that Yeadon was also placing himself in considerable peril, and squeezed his hand for one last time before pushing it away. Essentially he was telling him to leave, to return to the world of light and air and life, to save himself and leave Hasenmayer alone to die.

In a remarkable twist, as Yeadon returned distraught to the cave's entrance, Hasenmayer appeared behind him, having found a way out of the constriction in one final, desperate effort. He refused to show the other divers his air gauges, and the general opinion is that he had seconds left by the time he surfaced from within the cave. From that point on, the dark gap in the wall in the depths of Keld Head has been known as Dead Man's Handshake – a monument to the fortitude and courage of Hasenmayer and Yeadon in that moment of torment in 1978.

My fellow speaker at the RGS event was a man of this ilk: no nerves, a latter-day pioneer. Before the talk, however, he was an ashen, trembling wreck. During the course of the presentation, his slides stuck, resulting in him standing, stricken with fear and horror, staring miserably at 800 expectant faces. When I questioned him about it afterwards, as he fled back to his own subterranean existence, he looked at me with hollow eyes and simply muttered, 'I just don't know how you do that, I just don't understand how that can be enjoyable to anyone' – blissfully unaware of the irony of these words from a cave-diver.

Put me alone, in front of a hostile audience of a thousand, wearing no trousers, with a lecture to give about why Idi Amin was actually a deeply attractive human being, and I'll be perfectly happy. Put me at the foot of a mild rock face, in the hands of a fabulously well-qualified and experienced mountain guide, and you'll see a man wrestling helplessly with an army of demons.

*

The rope finally went taut, and Mike's voice drifted down from on high. 'OK, Mr Bonnington, ready to climb?'

I set out grimly, the rope coming taught as Mike took up the slack. Reaching upwards, I grabbed a large outcrop of rock, stepped upwards and broke several promises to myself by once again departing terra firma.

The ground was soon swallowed by the mist. I might as well have been on the north face of the Eiger for all the visual references I had. I tried to tell myself that it could also be a gentle scramble up a ledge in my local park in Bristol, but the mischievous forces of adrenalin and instinct made the thought immediately evaporate, to be replaced by the undeniable truth that I was on the Inaccessible Pinnacle, a feature that stuck out of the top of the ridge like a tombstone.

As I climbed towards Mike, I moved over the top of the edge of the pinnacle and straight into the wind that was being whipped along the curved face of the cliff. This carried tendrils of mist past me, a flowing river of cloud that swirled and eddied around my lonely form. Clambering grimly upwards, I soon saw Mike sitting comfortably on the edge of the face, calmly feeding the rope through his hands, always staying in contact with me, akin to playing a giant trout up a vertical stream.

'Well done, mate,' he grinned as I joined him. 'That was a master class in climbing.' He was plainly enjoying himself immensely, the complete bastard. 'Just hang on there, I'll sort out the next pitch.' He remarked to me later that I seemed quite relaxed as he prepared for the next section of the climb. I assured him that this was a simple acceptance of certain death and that if he had attempted to prise my steely grip from the rock as I stood there very bad things indeed would have happened to him almost immediately.

The next – and final – pitch was considerably easier, again completed by Mike going ahead, a rock athlete skipping up the face, flowing over ridges and cracks, moving ever upwards. I followed, a crabbing, cloven-hoofed, misshapen figure appearing Gollum-like out of the gloom to crouch once again by Mike as he prepared for our descent.

He turned to me with genuine warmth as he retrieved the last length of rope.

'That's it, mate – well done. You've conquered the Inaccessible Pinnacle.'

He stuck out a hand, which I managed to shake, having removed my white-knuckled digits from the rock next to me. As if on cue, the mist briefly lifted, carried away by a particularly savage gust. Alongside us the ridge fell away, a waterfall of rock, boulder and grass, crashing into the valley floor beneath, the stream shining in the sun as it went about its endless task of carving and sculpting the valley floor. Far from the surge of exposure and fear I had expected, I actually felt a rising euphoria. Climbers seem to me a strange breed, as I suppose divers do to them, but for a second their endless pursuit for the thin air of altitude made perfect sense, affording as it does a box seat, if only for a moment, alongside the gods.

*

They arrived gradually, announcing their presence by the occasional odd pinprick when I was working in the still air within the fank walls, followed immediately by a deep itch which I would scratch absent-mindedly as I continued with my chores. This was merely the advance guard, one or two individuals emerging slightly too early in the year.

Their moment arrived in the middle of July, when the grass was still slick after a heavy rain shower, and yet the air carried a hint of the warm day ahead. As I opened the bothy door in the morning, delighted to be emerging into the fug of a lazy summer's day, they rose to meet me.

The midge is the modern scourge of the west coast. Call them what you will – no see 'ums, the Scottish Air Force – when the conditions are right they appear in voracious clouds, homing in on the breath of any creature large enough to produce carbon dioxide, zigzagging along an invisible trail until close enough to settle and feed.

The word midge comes from the Gaelic *meandh chuileg,* translating as 'little fly'. Coincidentally, 'meandh chuileg' was precisely the kind of noise I made as I charged deliriously around the fank, arms waving about my head like some crazed simpleton, alternatively raging then sobbing at the relentless assault of these humming demonic squadrons. Although there are forty species of midge in Scotland, there are only five that bite people. It's always the females that bite, requiring blood before they can reproduce. At this point, there are all sorts of tired analogies I could draw from some of my previous relationships, but sometimes nature presents such a neat metaphor that it doesn't need a great deal of elaboration.

Of all the species of midge, there is one that rises above the rest, and that is the Highland Midge – *Culicoides impunctatus.* This whirring speck of pure malice costs the Scottish economy an estimated £280 million a year in lost tourist revenue, with wild-eyed and itching visitors fleeing south with vows never to return. Various studies have been requisitioned in order to estimate midge numbers, with figures ranging from 10 to 50 million midges per hectare of ground. The exact number is fairly irrelevant in my opinion, as once the figure passes a million or so life becomes pretty intolerable. The nearest true estimate of how grim it can be came from a study several years ago, which fastidiously noted that 40,000 midges landed in a single hour on a single arm. (The study doesn't mention what the other arm was up to, although some sort of vigorous scratching was probably in order.)

There is another revered and reviled midge: *Culicoides newsteadii* – the salt marsh midge. This was spoken of in hushed tones, referred to as 'the Beast of Arrochar'. Judging by the stories the locals told me, it was the size of a fruit bat, had a Black and Decker drill bit on its nose, and could drain the torso of a small boy in a single bite. Thankfully, we never crossed paths.

Feeding the stock was now done at a sprint, with food hurled over the wall and water slopped into troughs as I charged past. I would return to the extension wide-eyed, face reddened by self administered slaps, hundreds of tiny carcasses dotting my features. Antje and I would eat our breakfast under siege, watching the midges crawl and hum against the window pane, falling to the sill in a dark mass. It was torture on every level – the maddening hum of the ravenous swarms, the pain of the bites, the deep itch afterwards, and the lack of places to hide without being followed by their tiny, vicious, vindictive little forms. Squashing them was briefly satisfying, feeling like a small victory in an endless war, but each tiny form was replaced by a hundred others.

Doris became – not unreasonably – even more foul-tempered, moving around the enclosure in a mist of biting, swirling forms. She looked somehow pixelated (or pigselated), a dim figure in a bloodthirsty fog seeking any avenue of escape. She and Gemma would burrow deep into their shelter, piling bracken and mud upon themselves so only their noses were showing, sighing and twitching in their torment.

During the warmer weekends, the beach would see a reasonable number of visitors – never becoming excessively busy, but nonetheless playing host to small groups or families enjoying a day out. As the midges were most active at dusk, the end of any particularly still day always reminded me of the scene in *Jaws* when a fin is sighted in the waters off a packed beach. The resultant stampede, with wailing children snatched from the shallows and headlong panic, bore an uncanny resemblance to the moment the midges decided to swing into action, attacking flailing tourists in squadrons of immeasurable numbers.

For the deer in the hills at the head of the beach, life must have been intolerable, and I would occasionally see them lying in the water as the tide crept up the white sand, desperate for relief. Their ears would flick, their eyes would close, and they would lower their heads almost to the sea's surface, a giant brought to the edge of madness by a microscopic aberration of nature.

One terrible morning, when the midges were feeding in impossibly dense clouds, I emerged from the bothy, swathed in netting and shining with Skin So Soft, to see a tent in the dunes. The occupants must have arrived the night before, in the hours of darkness when the midges were at their least active. In due course, the door was unzipped and an attractive, bright-eyed, smiling young couple emerged to what appeared to be the first morning of their holiday. They were wearing, to my wide-eyed horror, shorts and T-shirts.

I wanted to shout, I really did. And yet, before the words 'Run, save yourselves, forget the tent you can buy another one' had passed my lips, the first midges had arrived. I started forward to warn them and then halted, aware that a large man running towards them wearing a burqa of dark muslin and bellowing oaths may be something of an alarming sight on the first morning of your romantic break. As their first slap was administered – the girl swatting her forearm absent-mindedly – millions of tiny wings blurred into life, and the assault began.

The young couple, so clean-limbed and innocent in the gentle morning light, went from a picture of serenity to a crazed, Morris-dancing, flailing duo within moments. The girl hopped up and down waving her arms around her head, her shrick carrying to me clearly in the still air. The boy – plainly a decisive young man with a cool head – immediately started to take the tent down, pausing occasionally to beat a rapid tattoo from head to toe. My final glimpse of them was their ungainly sprint over the dunes, the man with a mass of tent and poles under one arm, the girl shouldering two rucksacks and fumbling for the car keys. After they had dropped out of sight, I could still hear them, the beginnings of an argument rising in her shrill accusation, with his muffled response sounding rather subdued and bewildered, a man whose romantic break had turned into a nightmare.

Only when a mild breeze rose up or the rain drummed or the sun glowered was there any relief. The midges would duly sink back to the ground and wait. Their ideal conditions were geared perfectly for the temperate west coast, with overcast, cloudy, still days the times when they reigned supreme, the unchallenged lords of the flies.

The midge is an extremely interesting animal in terms of

the history of the west coast. The first round of Highland clearances took place to provide land for sheep, small farms and crofts becoming untenable as the need for increased rents caused landowners to evict long-standing tenants. This could be said to be the historic reason why the west coast of Scotland became a sparsely populated wilderness. The midge, however, has contributed heavily to it remaining that way. The link with the sheep is strong. Here is the animal that created the need for one set of clearances, and yet there is strong evidence that their heavy grazing removed many of the midge's natural predators, leading to a population explosion. The humble sheep has a great deal to answer for in the Highlands – perhaps this is one of the reasons they always look so shifty (as they should do, quite frankly).

It is only with the first frosts that the midge dies out for that year, their short life cycle completed, their mission accomplished. Perhaps it could be said that the midge performs a service, keeping the Highlands wild and untamed, unsullied by hordes of visitors. A recent study into their complete eradication foundered on the rocks of conservation, the argument being that they are a key part of the ecosystem. For me they represent nature run amok, a pestilence that prevents development and stifles income, perhaps an uncharacteristic slip of the hand when mother nature was crafting the masterpiece that is the west coast.

26 July

The Highland Games

There is a long tradition of games in the Highlands, stemming from a time when the clans would gather in a show of unity. Tests of speed, strength and endurance would be undertaken, with the winners of the heavy events becoming bodyguards to the clan chief, and the fleet of foot becoming messengers in battle. It was, and is, a celebration of Highland tradition, a joyous occasion for a proud people at the most prosperous time of the year.

The committee for the games organisation was the unlikely pairing of Linda and Alfie. Anxious to help out, I sought out Linda over a coffee in the Potting Shed.

She looked absolutely delighted – if slightly taken aback – when I offered to help. 'A real-life volunteer,' she said. 'We haven't had one of those for years.'

It was a lovely day, one of several over the previous couple of weeks that announced the re-emergence of summer. As the steam rose from our coffee and Peter and Jackie bustled about in the Walled Garden outside the window, Linda told me more about the games.

'This isn't really a very traditional event, it's more a fun get-together for the village. We've got all sorts of events; some are the classics; some are slightly more quirky. It's a case of all hands to the pumps, though. If people didn't come up to help us out, the games simply wouldn't happen. It's a bit like a garden fête gone mad. There is one thing that you simply can't miss, though – an annual classic that we all look forward to with bated breath. That part's a secret, though – you'll have to wait and see what it is.'

I asked her what event she would like me to organise, rather hoping for a Highland classic – perhaps some caber-tossing, the hill race, or the hammer.

'Right,' she said, tapping a pencil against immaculate teeth as she scanned a clipboard. 'You can organise the sumo wrestling.'

It was at this point I realised that perhaps this was not going to be the riot of tartan, pipe bands and intractable tradition that I had originally thought.

*

The day before the games dawned bright and breezy, important for the vast team effort required to set everything up. Turning up at the field, I could see a small army of helpers hefting marquee poles, putting up stalls, laying out tracks, or simply running round with cups of tea.

I found Linda in the midst of the throng, gliding from group to group in white, knee-length boots with high heels, tiny denim shorts and a halter top. I could also see Alfie in the middle distance, who thankfully had left his tiny shorts and high heels at home, plumping instead for solid work boots and overalls.

'Ah, there you are, Monty boy – fashionably late as usual. You can give us hand with the marquee.'

This was reminiscent of the barn scene in *Witness*, where the Amish community works in harmony to create a structure amid the American plains – the only things missing were the unusual beards and clashing steins of lemonade. The marquee gradually emerged, amid much bellowing from Alfie and sharp banter aimed – it appeared – almost entirely at me and my poor peg-hammering skills. A brisk squall appeared from the bay, spiralling over the brow of the hill and entering the semi-built structure, which at this stage looked precisely like a wind tunnel. The marquee duly tried to take off, with the extra manpower drafted in from the surrounding camp site the only thing that saved the day.

At the end of a long afternoon, the scene was finally set, with rows of neat, white tents clustered around the main marquee.

'You see, Monty boy, if you build it they will come,' said a perspiring Alfie. 'Has anyone mentioned the special event that you simply have to see? Well, obviously I can't tell you just yet, but trust me, you won't be disappointed.'

The next morning, under blue skies and sunshine, the first crowds began to gather. The games were taking place on the main field by the campsite, offering as it did a large flat area, lined by tall beech trees and affording a beautiful view down to the bay. The air hummed with insects meandering from flower to flower in the low shrubs that grew by the field, bright flags fluttered in the gentle breeze rising from the sea, and the marquee glowed under bright skies.

Stallholders had already begun to set up for the games, with local handicrafts in abundance, as well as the more unusual Applecross-themed stalls. I passed Peter and Jackie as they set up the Dog With The Waggiest Tail competition, and said hello to Andy as he arranged the Hammer The Nail In As Fast As Possible stall ('I'll just be using my nail gun at the end of the day – should see me walk away with the grand prize,' he said with a smile). Kayak Mike was organising the Hill race, and waggled a pen at me as I passed, raising a questioning eyebrow. I glanced up at the distant flag fluttering in the wind on top of the glen, and did my best to ignore him, the horrors of Glamaig still in my trembling thighs.

The main marquee was alive with brightly woven tartans, traditional carvings, atmospheric photographs and rows of cakes. The local ladies had set about the task with gusto, with creaking tables heaving with whirls of icing, bright yellow alongside strawberry red, with lavish swirls of chocolate sitting atop layers of rich cream and dark sponge. The air was filled with sweet smells, fresh grass mixing with warm bread, with the ever-present whiff of the canvas bringing the memory of long-forgotten camping trips.

Big Willie and Alfie were adding the final touches to the heavy events, laying out a series of poles to mark a throwing area. This was the one area of the games that remained entirely faithful to tradition, and I glanced with some alarm at the lumpen weights at the start line, remembering my distant promise to Alfie that I would be hurling and heaving with the best of them. Alfie saw me pass, and dropped one eyelid in a solemn and slow wink to show that he too hadn't forgotten.

By now the field was full of holidaymakers and villagers, with Linda darting around, herding lost stallholders into place, all the while fielding enquiries and making last-minute adjustments to stalls and stands. At last, with a bellow from their magnificently attired leader, the Skye pipe band struck up with a rousing tune and the games were under way.

Running the sumo-wrestling stall certainly proved to be one of the more enjoyable aspects of the day, and I could immediately see why it was such a strong pillar of Highland tradition (as I told one bewildered American tourist). Essentially it involved dressing two perspiring opponents in gigantic latex suits, forcing them to posture and pose at either edge of the ring, then giving them free rein to charge at one another. Each fight invariably ended up with one – or frequently both – would-be warriors lying helpless on their backs, legs and arms waggling comically as they tried to right themselves. There was one unfortunate moment when a competitive young man charged at his girlfriend and – somehow bypassing several layers of foam and padding – managed to inadvertently butt her on the bridge of her delicate elfin nose. I saw them the next day in the pub, with him staring sheepishly into his pint whilst she glared at him over the top of a large white plaster, offset neatly by eye shadow that consisted of two plum-coloured bruises.

Mike set off the hill-runners, and I watched them bound off into the woods like startled fawns. I had not the slightest hint of regret that I was not amongst them. I had an outstanding excuse anyway, as the heavy events were about to begin.

The heavy events are steeped in history and tradition. They consist of the shot-put, the hammer, the caber and the weight over the bar. The latter involves throwing a weight over a bar – I trust the naming committee for that particular event met late on a Friday afternoon. There is also an event at certain Highland gatherings called throwing the heavy stone, which involves throwing a heavy stone. There are many, many admirable aspects to Highland culture, but I'm not sure that imaginative names for games is one of them.

First for our little gathering was the shot-put. My fellow competitors were a mixture of swarthy locals and mildly inebriated tourists. There was a certain level of banter as we warmed up prior to the competition, with one muscular, kilt-clad fisherman from Lochcarron particularly vocal. I heard him mutter, with only a slight trace of irony, 'If I lose to a Sassenach called Monty, I may well have to tek my own life.' Amid ribald cheering and ironic applause, the shot was duly putted with minimum technique and maximum effort. In front of a well-oiled crowd, the competitors were spurred into great efforts, the air thick with beer fumes and guttural grunts.

My turn duly arrived, and I moved towards the shot as it lay on the grass. Local fishermen Big Willie, Donald (known to all and sundry as Gutsy) and of course Alfie were all involved in running the event, and met my arrival with bellows of encouragement.

'Come on, Monty boy,' shouted Alfie, 'show us what you're made of.'

I picked up the shot with a grunt, turned, nestled it under my chin in the approved manner, and without further ado threw it into the middle distance.

The most extraordinary thing happened – the shot sailed into the air, a weighty parabola in the sunlight. It flew and flew and flew, finally landing with a percussive thump. Perhaps I have always had a hidden talent for pointlessly throwing large lumps of metal onto grassy fields and never knew it. Perhaps I got lucky. Perhaps I was the only sober one present. Whatever it was, I had won the shot-put. The crowd applauded, Big Willie came and slapped me mightily on the back, and the kilt-clad fisherman looked suitably horrified.

My sense of triumph and the sneaky suspicion that I might be a mighty Highland warrior were duly quashed, however, during the next event – throwing the weight over the bar. I simply couldn't shift the idea that the 58lbs (26kg) of metal I was attempting to hoy over the bar behind me was going to land on my skull on the way down, crushing it like an eggshell.

'It's impossible for it to land on your head,' said the watching Alfie. 'It simply never happens.'

'Except for that one time, of course,' said Big Willie.

'Oh, aye, there was that one time,' said Alfie, looking faintly shifty.

I came a comfortable last.

Next came the caber, that most traditional of Highland sports. I was delighted to see that the Applecross version – far from being a mighty piece of Scottish timber – was more akin to a large snooker cue. This meant that everyone had a reasonable chance of tossing it. The exact definition of a 'toss' is the caber landing on its tip then falling away from the tosser (as it were), not towards him. We all had a tremendous time, with one slightly small and rather drunk contestant weaving and wandering his way through the field for several minutes, the caber swaying above him as Alfie bellowed at him to run for his life or be crushed.

The final event was the tug of war. I watched Antje first, as she had been press-ganged into the Applecross Women's team. Undefeated in living memory, the team was anchored by the stout and glowering Carolin, sister of landlady Judy. She braced heavy boots into the soil, took a turn of the rope around her wrist, and roared defiance at the opposition. She continued to do this as the teams changed ends, causing one or two of the ladies in the tourist teams to go quite ashen. Victory was a formality, with numerous high fives and rude gestures at the crestfallen opposition.

I congratulated Carolin for her outstanding technique.

'Yes, that was rather good, wasn't it,' she said, pushing her glasses up on the bridge of her nose. 'Now, you need to get out there and do it on behalf of the men. Don't let me down now.'

I was slightly alarmed to note as we took our place on the rope that the Applecross Men's team was distinctly on the lightweight side. Our first pull was against a group of bikers who had turned up specifically for the games, all large boots and tattoos.

'Don't worry,' said Kayak Mike, noticing my concern. 'You wait until you see our anchor.'

As the words still hung in the air, there was a movement in the crowd, a parting of the waves to allow a vast presence through. Bob – our anchor – duly appeared. He was slightly more of the pie-orientated school of training than a powerful athlete, but his gently wheezing presence immediately doubled the weight of our squad. He strode to the end of the rope, wrapped it round his colossal waist, leant back, took a last pull on his pint, cast aside the plastic glass, burped, and waved regally that the pull could begin.

We won our first and second bouts, with Bob leaning back and looking faintly bored as the opposition heaved and scrabbled at the other end, purple-faced and sweating. We were through to the final. Victory seemed inevitable.

The final, against an alarmingly well-drilled team visiting the games specifically for the tug of war, was going according to plan until disaster struck. The combined impact of the previous pulls and several pints of lager created serious gyro issues within Bob's vast frame, and he fell over. Bereft of our anchor, we were very swiftly going in the wrong direction, with Bob being dragged through the grass in our wake. The resultant wedgie, administered by eight men dragging Bob along the ground and the presence of several yards of bottom crack into which his shorts and pants vanished completely, left the crowd applauding thunderously. Although we went on to lose the last pull, I couldn't help feeling that we should have been awarded a special prize for sheer entertainment alone.

The games drew to their conclusion, with winners applauded to the skies – although the woman who won the Owner Who Looks Most Like Their Dog competition looked somewhat disturbed at the result. The crowds began drifting away, tired toddlers nestling on the sunburnt shoulders of their parents. The evening was drawing in – but by no means was the day over. That night the marquee was to play host to the annual dance, an event of legendary intensity. I helped Linda and Alfie as they packed away the stallholders' tables, clearing a space for the festivities to come.

'The big occasion draws ever closer, eh, Alfie?' said Linda, precisely loud enough for me to hear.

'Oh aye, I certainly wouldn't be anywhere else tonight,' said a tired-looking Alfie, pausing to mop his shining brow.

I finally buckled under the stress of several days of being baited by the locals about the great event. 'What? What is it? You have to tell me,' I spluttered, to the evident delight of both of them.

'Well, you'll just have to wait and see, won't you?' said Linda primly.

'Aye, stop being so impatient,' added Alfie triumphantly. 'That's the problem with your generation – it's all about instant gratification.'

The sun slowly drifted towards the tree tops, and the temperature dropped as the first revellers appeared, many resplendent in Highland dress. The band began to warm up, the beer flowed, and soon there was barely room to move in the fuggy atmosphere of the tent. The band was called Rhythm and Reel, and consisted of the standard instruments one would normally associate with any band working the party circuit, with ageing rockers playing a bass guitar, lead guitar and drums. There were, however, two notable additions. Two girls were playing a fiddle and a set of pipes respectively, both with long, dark hair and the flashing green eyes of the true Celt.

As the band struck up, the fiddle and pipes created a completely irresistible combination, rising and pulsing over the main beat. Maybe it was the strength of the beer, maybe it was the attractiveness of the ladies, but I swiftly decided that Rhythm and Reel were simply the greatest band I had ever heard. There wasn't really a dance floor, just a large section of the crowd that began to bounce and spin to the music, the pipes ricocheting around the confined walls of the marquee, energising tired limbs and demanding that everyone dance, regardless of age, fitness or sense of timing. Modern songs were combined with traditional ceilidh tunes, with the occasional gross approximation of organised dances taking place, locals and tourists joining together to spin and jig across the marquee.

Just when I thought the evening simply couldn't get any better, Mike nudged me and bellowed into my ear over the din, 'Here it comes, certainly the highlight of my year.'

In the midst of the throng, roared on by a delighted crowd, the considerable frame of Donald was moving towards the central pole supporting the marquee. Gripping it firmly, he draped one massive thigh around its girth, and began to pole dance. I glanced round to see Linda whooping and cheering, applauding Gutsy wildly. Andy and his wife Heather were roaring with laughter, and even Alfie looked around the door to smile and give a thumbs-up before returning to his money-collecting duties. The crowd were in his thrall, watching the show with expressions ranging from the slightly stunned through to the wildly appreciative.

Donald finally took a bow, returning to his pint with a triumphant wave. With the final great event complete, the Applecross Games could now officially draw to a close.

11

Last Days

In 1945, Gavin Maxwell set up a basking shark fishery on the island of Soay, an experience he recorded in his book *Harpoon at a Venture*. Maxwell battled seemingly insurmountable odds to try to make the fishery a viable concern. His first obstacle was attempting to harpoon and capture the sharks themselves, no easy task considering that this is the second largest fish in the oceans, up to nine metres long and weighing seven tonnes. This is the equivalent of shooting a harpoon into four elephants at once, fighting them from a small boat, then dragging them up the rocky shore of the island for processing. It was pioneering stuff – no one had ever attempted to commercially hunt the sharks, considering them too massive to be caught. His book reads as a monument to the human spirit, with his unquenchable resolve overcoming huge obstacles on a daily basis. The fishery was just beginning to operate efficiently when the money ran out, with the ruins still standing on the island, a mute testimony to failed endeavour.

It may seem incongruous that Maxwell, with his love of nature, set out to create a fishery. This was a very different age, however, a factor that comes through in his writing. He showed a remarkable empathy with the ecosystem around him on dry land; however, he regarded the marine environment with slightly more suspicion. The most notable example of this is his description of orcas off Sandaig, with Maxwell plainly terrified at their appearance, writing vividly of his headlong retreat back to shore when they appeared alongside his small rowing boat. He was similarly awestruck when he saw his first basking shark, a sensation shared by mariners throughout the ages. The poet Norma MacCaig was moved to write:

That once I met, on a sea
tin-tacked with rain,
That room-sized monster
with a matchbox brain.

Maxwell learned to respect the power and grandeur of the sharks, making meticulous notes on their natural history as the fishery progressed. Much of the knowledge we have today about the basking shark stems from his fastidious observations.

I had been hoping to see the great dark triangle of the fins from the bothy, and sat in the extension for hour after hour, scanning the channel. Although I saw plenty of activity, with seabirds galore, the occasional passing pod of dolphins and immense mackerel shoals ruffling the water's surface as they fed on dense clouds of plankton, I didn't see any sharks. It seemed to me that certain species of seabirds passed in precise formation, drawing my eye as my heart skipped a beat, two dark shapes moving slowly along the surface, a thrilling distance apart, plainly representing a monster shark beneath. It was only when one of the fins flapped its wings that the shape of two rather smug looking birds would clearly take shape, and I would sink back into my chair muttering darkly. Shags and guillemots were the main culprits, species that plainly have a well-developed sense of theatre when it comes to winding up wannabe shark observers.

The local fishermen advised me that it had been a poor season for sharks, with fewer sightings than normal and relatively infrequent aggregations of plankton. This may have been due to the terrible weather throughout June, or simply that the sharks were feeding elsewhere. Nonetheless, I remained determined that I would encounter one before my time in Scotland ran out. The word on the wind was that there had been numerous sightings off the south-eastern tip of Skye, so one fine August day I decided to take to the road.

My fanatical desire to see a basking shark was hardly a new phenomenon. Around our shores there is probably no animal that has created quite as much interest. During Victorian times, the reeking remains of sharks would occasionally be washed ashore, creating claims of sea monsters amongst the mildly hysterical. This was not as bizarre as it might initially sound. For a start, there is the immense size of the sharks, combined with the fact that the lower lobe of the tail and the gill arches are one of the first things to rot away when they decompose. This leaves a very passable imitation of a sea monster indeed, and numerous sepia photographs exist of frock-coated, bewhiskered men looking stern beside festering carcasses, eyes watering furiously – rotting shark smells like nothing else on Earth.

Even modern science remains mesmerised by the sharks. A huge question remained unanswered for many years: where did they go in the winter? The dark shapes that cruised the shallows throughout the summer would simply vanish like ghosts as winter set in, sculling into the depths during September to reappear the following spring. Some truly bizarre theories abounded, my personal favourite being that they would sink to the sea floor to hibernate whilst buried in the deep silt. It is only in the last few years that this question has finally been answered – the sharks simply move offshore, feeding on deeper aggregations of plankton.

The reappearance of the basking sharks around our shores occurs in late spring down in the south of England, and towards the end of the summer up north. This simply follows the patterns of the plankton blooms, and allows for another more recent phenomenon to occur on an annual basis on the south coast. Here, regular as clockwork, a tourist will spot a massive fin cruising off shore, take a picture of it on a sequin-encrusted mobile phone, and the national press will temporarily lose its mind. In the year that I was in the bothy, my favourite headline, from a particularly respectable national newspaper, shrieked 'Killer Sharks Swim Off Our Shores', sending countless families out to buy Kevlar swimming trunks and bang sticks. The fuss invariably dies down very quickly, but it's a phase of the year I have begun to look forward to, providing as it does a thoroughly enjoyable interlude to any no-news day.

In my personal quest for the sharks, I had a secret weapon of considerable dimensions in the form of my friend Dan Burton. I had known Dan for many years through the diving circuit, and we had been on a number of expeditions together. He was utterly fearless, and would dive with a ragged assemblage of webbing straps and wheezing valves, clicking away with a camera encased in a home-made housing which he had knocked together in the shed in his back garden. His photos were world class, and he had created a thriving business for himself over the previous decade.

Dan had taken up paramotoring several years previously with the concept of strapping a lawnmower engine onto his back to whirr his way through the sky under a billowing canopy proving completely irresistible. In the early days of taking up the sport, he crashed on a regular basis: 'I'm at about fifty per cent in terms of crashes to landings,' he told me proudly one afternoon as he glued together another section of the paramotor after one his more exuberant touchdowns. I personally witnessed one particularly exciting take-off, watching him lift his legs as he roared over a barbeque at a camp site, the ashen-faced cook staring at his imminent demise with a smoking burger held motionless in a pair of tongs, only waving it vaguely in passing as Dan bellowed 'Sorry!' over one shoulder and clattered towards the horizon.

Dan had just flown from Land's End to John o'Groats as part of a small team, becoming one of the first ever to fly the length of Britain in a paramotor. When I met up with him, he was still red and raw from his time in the air, with flakes of skin curling like woodshavings from his cheeks, the result of hours of exposure to wind and rain. He had also grown a ginger beard, which clashed violently with his ruddy skin. The overall effect was somewhat alarming.

'It was bloody marvellous,' he bellowed to me in the pub. 'I dropped a thousand feet [300 metres] over a mountain when my chute collapsed, nearly landed in a marquee, and got blown up a valley in a storm. You really should give this a go, you know.' (I vowed internally to never, ever leave the ground again.)

Dan is a force of nature, a creation from a Ripping Yarns story. Chat to him for more than ten minutes and you swiftly realise why Britain once had an empire. He was coming along to be my eye in the sky as we hunted for the basking sharks, flying along the rugged shore of Skye.

Finding marine animals depends on a number of key factors. The first is encountering the right conditions – weather, waves, wind and the presence of food. The second is the help of local fishermen and boat skippers – there is simply no substitute for the sort of deep-rooted knowledge that is only acquired over generations. Next comes patience – the open ocean is not a zoo, and sometimes the only thing to do is wait, no matter how frustrating or uncomfortable that may be. The final – and most important factor – is luck. One needs the fates to conspire to bring you and the animal together, crossing paths for a moment in the immensity of the sea.

We travelled to Skye and booked into a small hotel. The next morning, as we prepared to launch the RIB in the picturesque harbour of Armadale, we certainly seemed to have the right conditions. The sun shone from a sky streaked with wisps of high-altitude cloud, a mackerel sky over the deep blue of the water beyond the headland. There were also strong easterly winds, driving the waves towards the steep sides of the cliffs leading to the point. There had been several days of sunshine, creating a warmer upper layer of water sitting on a colder layer beneath. This essentially traps the basking sharks' food close to the surface, creating an all-you-can-eat buffet through which the sharks cruise, massive mouths agape. (The inside of the shark's mouth, which can measure a metre across, is white. This is thought to be because the plankton is light sensitive and is drawn to the interior of the mouth, actively sculling to their own demise.)

Peter Fowler was our local contact. As he was the man who had sold me the RIB and its entertainingly quirky trailer, I felt that at some level he owed me a favour. Since he ran the local whale-watching safaris, he would also be able to point us in the right direction. He cheerfully agreed to keep us posted as he ran his tours for that day.

Dan had scouted out a suitable beach for launching, and was soon tinkering with the paramotor on the pebbles. It was a fearsome-looking contraption, with a large propeller encased in a red cage, attached to a small engine, which in turn had a padded seat strapped to it. It looked like the sort of thing a primary-school class would create when let loose in a junk yard. The chute itself lay on the beach, a diaphanous mass of lines and silk, billowing and twitching in the morning breeze.

'I've heard of a chap whose straps came undone once,' said Dan unexpectedly. 'Simply fell out of the sky. Bad deal that.

Do you like my flying suit?' He rubbed the red material that was stretched over the beginnings of a paunch. 'I wore a very tight one specially for you.' With a smile, he turned back to the paramotor, muttering and tinkering – a man in his element.

Having completed his final checks, Dan hefted the entire contraption onto his back. It was time to launch. With a mighty roar of the engine, Dan set out down the beach, the wing pulsing and flexing overhead and his legs furiously pumping beneath in a spray of pebbles. For one terrible moment, I thought that he wouldn't get off the ground and would simply run into the sea, vanishing beneath the waves with a thrashing propeller and bubbling engine to mark his progress. At the very last second, however, he lifted off, skimming low over the water's surface and then rising into the blue sky to the soundtrack of a triumphant whirring of the propeller.

It was my turn to join him, and I sprinted to where the RIB bobbed in the shallows. Turning the bow out to sea, I could already make out Dan flying above the headland, scanning the waters beneath.

'Hello, Monts,' Dan's voice, distorted with static and the noise of the motor, crackled through the radio. 'Lots of activity out here. Big line of plankton with several shoals of mackerel working the surface – looks very promising. Get beneath me and you'll see for yourself.'

I gunned the engine and raced to the point below where Dan was slowly circling. Sure enough, the surface of the sea was ruffled by an immense shoal of mackerel, feeding on the plankton drawn to the surface by the warmer water. This was a very good sign, as the presence of the mackerel on the surface indicated that the plankton was shallow, giving us an excellent chance of sighting basking sharks. The only reason the sharks are seen is that they go where the food is, and if it is on the surface they will happily follow, presenting that heart-stopping sight of a magnificent dorsal fin knifing through the waves. The fishermen in Applecross had told me that for every single shark seen on the surface there are always two underneath. They also spoke in hushed tones of the massive aggregations witnessed only rarely, several hundred sharks drawn together to feed, a layer of bodies beneath the hulls of the boat like a dark fleet of submarines.

Dan whirred off to investigate another quiet bay, and as he did so Pete's excited voice crackled over the radio.

'Monty, two sharks in a bay just north of the Point of Sleat. I'm looking at them right now, and one of them is a whopper. It's a small bay with a natural arch on the shore, and lots of gulls overhead. I suggest you get over here.'

Dan was already turning in response to the call, with me following in his wake. The bay was a few hundred metres away, and soon the radio squawked into life again: 'Mont, I can see it.' Even over the static and engine noise I could hear the excitement in his voice. 'It's bloody Moby Dick. Get over here quickly.'

Dan was spinning in tight circles over the bay, still two hundred metres ahead of me. When I was past the headland, I slowed the engine, and crept forward with the waves slapping the hull and the engine burbling behind me. I scanned the water ahead for the shape of the fin.

Dan's voice crackled over the radio. 'Mont, it's gone deep. I can't see it any more.' I felt a stab of disappointment. 'I'm going to pull back and let you drift onto it. My engine is also making some interesting noises. Might pop back to the beach and have a quick look. Good hunting.'

With a final wave, Dan wheeled off and set a course for the beach, a light smudge at the base of the dark green fields fringing Armadale in the middle distance.

I turned my attention back to the water around me, sitting static in the centre of the bay, rocking gently in the small waves. I was alone in a wild cove, just short of a headland that jutted into the mysterious waters off the Isle of Skye, with only the gulls and a massive, dark, sculling monster for company. The small boy within me held his breath, waiting for the fin to break the surface.

I scanned the waters constantly around the boat, eyes sweeping from the stern round the bow, then across the other side. As I completed one circle, I came round again to the same small patch of sea, and there, rising like a great dark sail only metres from the boat, was the fin of the basking shark.

It didn't look real, being altogether too massive to belong to any animal, regardless of size. It was slick with water and as black as newly laid tar, a sinister glossy triangle that jutted through the surface, swaying as the shark moved forward, mouth agape. Several metres behind, the tip of the tail fin eased from side to side, driving the body through the green soup of the plankton. I shouted involuntarily, even though I was quite alone. The fin was a ludicrous, thrilling, magnificent sight, straight from a Hollywood film, cruising alongside my little boat.

As if this wasn't enough, the shark actually began to turn towards me. I could clearly make out the white cave of its mouth, opening and closing, gulping as it processed the seawater within, filtering it through the massive gill rakers. It passed directly under the bow as I hopped from foot to foot and laughed at the helm, its great dappled body passing slowly before me, gently sinking into deeper water as it became aware of the obstacle overhead. I tried to grab my snorkelling gear and camera kit, whilst simultaneously attempting to start the engine to follow the dark shape as it moved to the far side of the small bay. I took a moment to calm down, mask still dangling from one hand, and exhaled in a single long breath. The shadow before me was indeed huge, large even for a basking shark. It made much more sense for me to observe it from the surface before attempting to photograph it within the water. I wanted this encounter to last as long as possible, to enjoy my moment alone with this giant animal. Placing my snorkelling gear to one side, I sat back and gently pushed the throttle forward, bringing myself slowly up alongside the shark.

The water was relatively clear, and I could make out the entire length and bulk of the shark. The massive body was not uniform grey as I had initially thought, actually consisting of a series of long, brown streaks down the shark's flanks, mixed with varying shades, from light grey to almost black.

The nose poked through the surface, a bulbous cone leading the fleet of fins, hinting at the bulk of the animal beneath. Closer and ever closer, all sense of perception vanished as this vast animal filled my vision, with the first stirrings of fear rising within me. I was alone, in a small boat, with a gigantic shark close by. I had experienced similar sensations in the past when in the presence of a basking shark, the ancient portion of your brain never quite accepting that the grey monster before you is harmless.

I shadowed the shark for several hours, my sense of wonder unabated. It was late afternoon by the time I turned the boat back to land, gunning the engine, letting the wake thunder behind me and the wind roar past the bow. I rounded the headland to see Dan standing on the beach, tinkering – as ever – with his paramotor. He heard my outboard as I approached, turning to wave a greeting as I crunched into the shingle of the shoreline behind him.

'Did you see it?' he asked. 'Tell me you saw it.'

'I did indeed, a monster, a giant, a . . . ' I trailed off.

'I'd say it was a behemoth as well, with possibly a touch of leviathan thrown in.'

'Precisely! Thanks so much, Dan. You are indeed the man – you can up-diddly-up with the best of them.'

Dan shrugged modestly, and leaned forward to shake my hand, as delighted by my encounter as he had been about his own.

'You might like to see this by the way. Had a few propeller snags.' He returned to the paramotor, and rummaged amongst the constituent parts. 'Here's my propeller, or what's left of it.' He held up a shattered stump.

'Turns out that I hit the beach as I launched – snapped half my propeller off. Didn't realise though. Thought the old girl was vibrating a little – turns out the prop was trying very hard to tear itself apart throughout that flight. Quite exhilarating.'

It had been quite a day – something perhaps we both would have preferred to dress up as a serious exercise in observing and filming sharks. However, as I looked at Dan holding the shattered remains of his propeller, and past him to the RIB as it rocked gently in the small waves off the beach, it seemed to me that we were very much two small boys, one with a flying machine, one with a speedboat, tired and happy at the end of a day searching for monsters in the wild waters off a mystical island.

*

Dan left early the next morning, heading back home with a final cheery wave and a promise of adventures to come. For me, there was a long drive back to the bothy, preceded by the considerable administrative task of loading the RIB onto the trailer and packing the snorkelling and filming gear. I then wound for several hours through single-track roads, over the Bealach na Ba, and finally back down the long straight into Applecross, hollow-eyed and yawning as the boat rattled and groaned on the trailer behind me.

Antje was away for a fortnight, tagging penguins in South Africa, so Andy's wife Heather had agreed to keep an eye on the stock for me. As I turned the corner into Applecross bay, my mobile rang.

'Monty, it's Heather. Thank goodness you're back – your pigs have been having lots of adventures. You're quite the talk of the village.'

The pigs had finally tired of their enclosure, she told me, and had gone to work in earnest on the gate. As she described the rigours of what had been a long afternoon for her, repatriating pigs who were intoxicated with their first taste of freedom, I could just hear a catch in her voice, something akin to post-traumatic stress.

'I think I've got the gate back on, but you need to get down there and check. They certainly seem to have a taste for the outside world now, and I think they'll have another go.' She sounded faintly shell shocked. 'Good luck,' she added wearily.

Things looked normal as I approached the bothy, with the sheep grazing peacefully on the gentle rise of the hill, and the façade of the cottage looking as pretty as ever. Approaching the front door, I placed my ear against the wood, and heard a murmur from within, the merest hint of a contented grunt, with a tapping of small trotters on cobbles. I rattled the lock, and the noise stilled. The pigs, it seemed, had taken up residence in the bothy. This was not going to be pretty.

I edged open the door, and peered around it cautiously. The scene that greeted me was much like the aftermath of a Viking raid, with everything turned over, trampled, and scattered. The chairs were all on their sides, the table sat at a drunken angle, my camping kit had been thoroughly investigated, and the duvet was half off the bed. I craned my neck further, half-expecting to see both pigs sitting in front of the fire drinking my whisky and smoking my finest cigars. Instead I saw more chaos, with the books off the low shelves, and my fishing tackle scattered in the far corner. Of the pigs there was not a trace, although some snuffling from the yard told me that they had found the raised vegetable patch – their equivalent of an open salad bar.

I marched grimly through the bothy and into the yard, to be greeted with a squeal of delight by Gemma, who was standing in the raised vegetable beds. She looked – as ever – delighted to see me, with her tail swishing in excitement and her pointed feet giving the normal little Riverdance of delight. She had the remains of my last cabbage hanging out of her mouth, and a vast distended belly full of my turnips, lettuce, beetroot, and carrots.

Giving a small groan of horror, I turned my attention to the more serious matter of Doris. I had visions of her having escaped completely, wreaking havoc in the local campsites and crofts, a psychotic pig loose in the village. Walking through the yard, ignoring Gemma's grunts and squeals of pleasure at my passing, I turned the corner into the pigs' enclosure, the gate hanging off its hinges.

At this stage I thought things couldn't really get much worse. However, this ignored Reuben's immeasurable capacity to add his own unique twist to occasions like this. He had been following dutifully at my heels, and rounded the corner as I did. It was he who saw Doris first, and he who decided that now would be the perfect opportunity to have some fun with these strange creatures with whom he had shared his home for the last five months.

He bounded forward, tail circling furiously and ear pricked. Doris looked up, snout festooned with the remains of vegetables and roots, and squealed in anger. Reuben put the brakes on, timing his skidding arrival at Doris to coincide precisely with my bellow of concern. There were also a number of chickens in the enclosure, which had been chased by the pigs and taken flight over the wall of their pen, and they began to run around in the approved chicken manner – squawking and flapping and pointlessly attempting to take off.

Reuben heard my shout and dropped to the floor. Doris charged towards him, scattering panicking chickens in a blizzard of feathers and flapping. I was running in from the other direction, waving my arms and attempting to distract Doris before she got to Reuben.

We arrived at Reuben simultaneously, just in time for me to lift him off his feet and for Doris to deliver a well-aimed bite at his furry backside. He howled in outrage, and twisted in my arms, but I hung on grimly and ran to the back door of the bothy, bundling the dog through and slamming it shut behind me. I sprinted back to Doris, grabbing a bucket en route and throwing in a handful of feed. I rattled it at both pigs, who glanced up and instantly trotted towards me. (I have no idea what they put in feed pellets, but it seems to be the equivalent of a porcine class A drug.) They both followed me meekly back to their pen, where I threw some feed on the floor to occupy them, and began the job of restoring order around the croft.

After rounding up the chickens, closing the gate to their enclosure and scattering some feed on the floor, I left them pecking contentedly and scratching the soil. I retrieved Reuben from the bothy who wore a look of faint annoyance at having been bitten on the backside then thrown over the lower half of a stable door onto some rather hard cobbles. I then went back to the pigs' enclosure to figure out how they had escaped. The pigs had plainly given their plan a considerable amount of thought. Through some well-judged burrowing, barging and chewing, they had finally reached a point where the gate could be lifted off its hinges. I had always been impressed by the power of the massive muscles that bunched at the backs of their necks, the hydraulics to raise the shovel of the twitching gristle of the nose. The gate had been off in a moment, and the pigs had trotted triumphantly out. Bedlam, chaos and destruction followed eagerly in their wake.

I replaced the gate itself, hooking it back onto the hinges, then spent half an hour moving the biggest rocks I could against the lower rung on the enclosure side. Soon there was a barrier akin to a dry-stone wall erected by someone on mind-altering drugs, a haphazard collection of boulders piled higgledy-piggledy (and I now knew where the last word in that expression came from) atop one another. Finally I secured the gate on both sides with pieces of rope, which I lashed, turned, tightened, and then tugged one time more.

Throughout this entire process, the pigs appeared to be slumbering peacefully, emitting the odd contented grunt, their only movement the occasional flick of a hairy ear to drive off noisome midges.

They were, however, watching me.

I would glance up occasionally and catch Doris's gimlet eye trained on the gate, studying me through slitted lids. Keith had warned me never to let the pigs escape, as once they had a taste for the wide open spaces, they would become habitual escapees, forming tunnelling committees and arranging elaborate exercise sessions using a wooden horse. I had the distinct impression that careful notes were being made of every stage of the gate refurbishment process.

Pigs are, after all, remarkable animals. Perfectly adapted to forage and investigate, to adapt to most conditions and to savour the ultimate in opportunistic diets, it is no coincidence that they were present in every forest and beaten savannah before man put in an appearance. Far from being angry with the pigs, I was actually lost in admiration at their ability to problem-solve, to apply a mixture of intelligence and brawn to work their way out of their enclosure.

I was fairly confident, though, that their wandering days were over. Short of erecting searchlights and machine-gun nests, I couldn't really think of anything else I could do to keep them secure, and retired to the extension for a quiet cup of tea.

I was awoken the next morning by the unmistakable sound of contented grunting emanating from the vegetable patch. I clambered out of bed, and peeked round the door into the yard. Gemma squealed with pleasure, trotting towards me through the shattered remnants of the turnips and the twisted corpses of the carrots. Doris merely glanced up and grunted contemptuously.

Herding them once again back into the enclosure, I saw the gate had been methodically dismantled. First the rocks had been removed – a moment's work for both Gemma and Doris working in harmony – then the rope carefully chewed piece by piece. The final stage was the lifting of the gate again.

I finally managed to contain the pigs – after two more escapes – by piling rocks against the outside of the gate, wedging each between the bars. This meant that the gate weighed a few hundred kilos, with the wooden bars preventing them getting their noses under each rock. It didn't stop them trying though, and for several days I drifted off to sleep to the sound of the surf and the quiet murmur of what could only be heated planning sessions at the front edge of their enclosure.

*

Over the summer months in the bothy, one of the highlights of my day was the visit of the Royal stag that had taken up residence on the hills behind the beach. Every evening, I would glance expectantly out of the window, awaiting his measured progress down the dune and onto the golden sand, watching his antlers branch and develop over the weeks and months, spreading and lengthening from small buds into velvet-clad weapons of war. One evening, as I had crept out of the bothy in an attempt to get a clearer view in the soft light of dusk, he had raised his head and stared directly at me. Usually this was a sign that I should slink back into the extension. On this day, though, I decided on a whim to keep walking. This was towards the end of the summer, when his antlers were a full, magnificent twelve points, and the testosterone had just begun to course through his system. His gaze met mine as I approached, not stalking or crouching but standing tall and walking confidently. Instead of trotting back towards the dune, he turned to face me, planting his feet and presenting the full chamber of his chest, dun-coloured curls coating the massive spread of his neck, his dark eyes staring directly into mine. I was looking at a magnificent animal in his prime, standing exposed on a sweep of beach in the face of man, staring down his age-old enemy. He allowed me to approach to within ten metres, where I crouched as if in supplication, firing off photograph after photograph. After a few moments, he turned and began to walk parallel to me, moving past with a final glance to head towards the dark hills above the bay. I sat on the sand and watched him climb through the dark morass of the heather, touched by the proximity of the encounter and that final glimpse of his grace and power.

Such sentiments were not amiss in Applecross, where the stag was regarded as an essential part of the landscape, both geographically and economically. Hunting the deer was as much part of the history of Applecross as crofting and fishing, with the battle between stag and man running over thousands of years – one armed with intelligence and technology, the other with a battery of senses, wings on their heels, and an ability to blend effortlessly with the hills and glens. It remains an even contest to this day, with many a stalking expedition ending in the sight of a distant herd heading for the horizon, the approach of man betrayed by a flicker of the wind or the turning of a stone underfoot.

It was now late August, and the hills were turning to rust as the bracken browned and shrivelled. For the stags it was time to prepare for the fight. All would be sacrificed for sex, the beasts on the hill turning themselves into pocket battleships carrying a precious cargo of genes. The stakes were the highest possible, with eighty per cent of the young being sired by only twenty per cent of the stags – the remainder left to wander the hills. Contrary to popular belief, it was not simply the stag with the largest antlers that won the brutal contest of the rut. Bodyweight was all, the steep hills and glens being the realm of the heavyweight, the antlers merely a means of holding and twisting the adversary. The more experienced stags would use the terrain to good effect, with the best position being the uphill charge, adding momentum and gravity to 200 kilograms of snorting, pistoning muscle. The animals in poor condition would simply not reproduce, a genetic footnote to be picked off by natural selection.

With an irony that will not be lost on those who campaign for the abolition of hunting, the man who cared most passionately for the welfare of the deer was the local stalker Dave Abrahams – known as Dave Keeper. He was the epitome of the quiet man of the land, keeping his own council and possessing a deep, measured approach to all things. On most nights he could be found nursing a dram at the end of the bar, swirling the warmth of the contents, talking softly to one of the fishermen or his stalking clients for the day. His ability to sense impending drama, developed over twenty years out on the hill, meant that it was invariably Dave who stepped in at the first hint of a disturbance at any village event, gently laying a rough palm on the shoulder of an exuberant drunk, accompanied by a quiet word and a half-smile. He was at once the guardian, recovery driver, part-time veterinarian, mentor and counsellor for the village – regarded with affection and respect by all.

Dave had already taken me out shooting on the range one summer afternoon, the military man in me intrigued by the skills required to take down a stag at considerable distance in a swirling wind. I had been thoroughly rattled by his brooding presence. I wanted desperately to impress, and had shot badly, spraying bullets liberally around the target. Embarrassed, I wandered miserably back to the vehicles with him, searching for the right words to explain my incompetence.

'Dave,' I said, trying to make a joke of it, 'I feel like I should do the right thing and simply top myself. We've got the gun – why not give it to me right now and I can stay behind. All you'll hear is a lone gunshot.'

Dave pondered for a while, the end of his cigarette glowing as a representation of his thought process. Finally he spoke. 'Having just seen you shoot, I can't imagine it'll just be a single shot. I'll leave you a few more bullets just in case.'

'That's what I like about you, Dave. You're so splendidly stoic.'

This took even longer to elicit a response – almost half a ciggie. 'Now, Monty boy, if you're going to be using complicated words like that, this could get tricky. Not being too familiar with your public-school language, I'm wondering whether I should be taking my jacket off or shaking you warmly by the hand.'

I hastily assured him that stoic was a very good thing indeed.

I had chatted to Dave at length in the pub about the process of the stalk – not just the time on the hill itself, but the rationale behind it. As a man so profoundly in touch with the land, it struck me as incongruous that he spent a great deal of his time leading clients on the hunt, shooting the very animals he loved so much.

This took some time for Dave to process, and I could sense the turmoil within. Finally, looking not at me but into the amber depths of his glass, he answered.

'Monty, if I never had to shoot another animal in my life I'd be delighted. We do have a duty though, us that live on this headland, and that is to keep the ecosystem healthy. The ancient predators of the deer are gone – the wolves, the bears – perhaps a fox or an eagle will take the occasional fawn nowadays, but essentially we are now controlling their population. We are in the unfortunate position of playing god here – without us the population would explode, way beyond the capacity of the food available, and you would see misery on a huge scale. Even controlling the herds the way we do, I see some deer in appalling condition during the harder winters. Miserable-looking beasts shivering with heads hung low, awaiting death. It's better that we take out the weak animals in the hunt prior to the depths of winter – it really is.'

He glanced at his hands, fingers gently tapping the bar, plainly ill at ease to be speaking with such candour.

'We live in the most beautiful place on Earth, us Scots. We're custodians, that's all. It's up to us to look after it for the next generation – there'll be healthy deer on these hills long after I've gone, that's what I hope anyway.'

Having devoured many a venison steak in the pub and at the Walled Garden, I resolved that I too should go out on a stalk with Dave. It seemed particularly unfair to judge from afar, to walk the verdant glens so carefully managed by the Estate Trust, without exposing myself to the cull, to the moment of pulling a trigger with a deer in the crosshairs. I asked Dave if he would take me out for a day on the hill, and he readily agreed.

We met outside the deer larder next to the big house on the estate. The larder was the processing point for any deer shot during the stalk, and had a rather sinister feel, with stark, white walls and a chilly interior – all gleaming, stainless-steel sinks and wicked-looking hooks on shiny metal rails. Dave was waiting for me, puffing contentedly on the ubiquitous rolled cigarette and chatting to his assistant, or ghillie, Tim. I had met Tim several times during my time in the village, and he seemed a dynamic, cheery figure. He was a bundle of energy, a mobile collection of tics and twitches as though the colossal dynamo within needed a perpetual outlet otherwise some sort of physiological explosion would occur.

He walked towards me and extended a hand, shaking mine in a single firm downward motion. 'Morning, Monty – "To him whose elastic and vigorous thought keeps pace with the sun, the day is perpetual morning" – that's Thoreau, by the way. Should be a good day today.'

It dawned on me then that perhaps Tim wasn't your average ghillie. Dave merely raised his eyes skyward in the background, and took a particularly long pull on his cigarette.

The plan for the day was to head to the very top of the Bealach na Ba, strike out towards the stark ridge of the Sgurr a Chaorachain, then move down the massive shoulders of the hill towards the road that snaked around the loch at its base. Here Tim would meet us with the Argo, a fearsome eight-wheel drive vehicle used for recovering the body of the deer after the hunt. This sat in the trailer behind Dave's pick-up, looking much as if it should really have been trundling over a moonscape. It also looked to be about as much fun as it was possible to have on wheels.

Dave was studying the wisps of smoke emerging from the end of his cigarette. He waved me over.

'You may think this is nothing more than a bad habit, but let me tell you, this is a highly technical piece of kit,' he said, waggling the glowing end vaguely in my direction. 'You can see the wind is twisting and shifting all the time – could be a tricky day today. We'll give it a go anyway, but the odds favour the stag when the wind is like this. I may have to light another soon to really establish the best direction for the stalk.' He even managed to say this without a smile, which was a fairly good effort in the circumstances.

We piled into the vehicles for the twisting drive to the

top of the Bealach, the village dropping away beneath us, a white line of cottages huddled on the shore of the great golden curve of the bay. The incoming tide was creeping up the shallow slope of the sand, sliding up the land like mercury in the dull glow of the morning sun.

Climbing the Bealach was akin to moving across a vast swathe of a continent, clambering into the clouds leaving lush fields behind to be replaced by stark rocks and low grass, a wind-scorched Alpine landscape clinging to the edge of the Earth, home to eagles and deer where the laws of natural selection were brutally magnified. Stepping out of the vehicle, I had to turn up my collar and hunch my shoulders, stamping my feet to prepare for the day ahead.

Dave was dressed perfectly for the hill, his clothing a series of soft tweed, dun-coloured fabrics and supple leather. This entire assemblage was topped – naturally – by a deerstalker hat, tied by a neat bow that fluttered in the wind. He carried a stout walking stick, and across his shoulders lay the rifle, looking sinister and wicked even when sheathed in a dark canvas case.

Tim leant out of the car window to wish us luck.

"'Searching is half the fun: life is much more manageable when thought of as a scavenger hunt as opposed to a surprise party." That's a Jimmy Buffet quote, by the way.'

Dave sighed. 'Aye, that's very interesting. Now go to the low road and we'll see you there.'

Tim laughed and gave a wave as he moved away, my final glance into the car providing a snapshot of a book cover on the back seat – *Philosophy for the Common Man*.

Dave was already striding towards the rolling hills off the edge of the road, and I moved quickly to catch up. Even when he was only fifty metres ahead, he blended perfectly with the landscape, not simply a measure of the right clothing but an economy of movement that matched perfectly the rhythm of the high moor. Even now, before the stalk had truly commenced, I could see that he instinctively kept off the skyline, moving soundlessly through the small valleys, picking each step atop mossy rocks and firm tussocks.

After a few minutes of walking, I became a bit impatient with the slow progress. Despite my noble intentions to remain detached from a process about which I still harboured certain reservations, the prospect of the stalk, the pleasant ache of my muscles on the walk, and the wind that whispered of prey in the hills ahead had awoken basic instincts within. I could feel my heartbeat increase and my senses sharpen, perhaps the ancient hunter that lurks within us all finally stirring after a lifetime of dormancy.

Almost without realising it, I kept drawing level with Dave and occasionally forged ahead. He sensed my urgency and, on one of our short breaks sitting in the lee of the wind, he stared into the middle distance and – unprompted – began to speak.

'Let me tell you a story, Monty,' he said. 'There were two stags on a hill once – an old wise stag, and a younger beast full of energy and testosterone. They saw a group of hinds on the valley floor below, and the younger one said, "Let's gallop down there right now, and we could have some of them." The older stag thought for a while and replied, "Let's go when we're ready, walk down, and we could have the lot."'

I continued the walk behind Dave.

As we continued to climb onto the ridge of the hill, the

views beneath offered an image of a landscape untamed by the hand of man, a wilderness on mainland Britain that remained a fortress for wildlife. The ridge continued away from us, meandering into the clouds in a dragon's back series of twists and peaks. Down its shoulders tumbled scree and rock, charting the path of the glacial forces that had shaped and sculpted it so many millions of years before. The great spread of the hill and valley before me seemed to itself to breathe, the clouds scudding overhead at each exhalation, the loch shining in the valley floor hundreds of metres beneath, the green swards of grass and moss on its great shoulders themselves crossed by a venous network of glinting streams.

As we moved over the gentle curve of a col towards the high ridge, Dave pointed out three golden eagles high above. 'Two adults and a young one – flying training, I'd say.'

As he spoke, the eagles span and stooped, twisting in the air to present wickedly sharp talons to one another, riding invisible rapids and eddies, their distant calls carrying to us below. The sound was cupped and amplified by the great hills and valleys around us, becoming the very essence of the wilderness around me – the exultant shriek of eagles as they tumbled in flight.

There was a lifetime of knowledge in Dave's assessment of the ground ahead, an instinctive appraisal of the wind direction, the consistency of the next footfall, the potential position of the deer, and possible routes of approach should one be sighted. He was working on a combination of instinct and knowledge, following the same deep gullies and tumbling streams that man had for centuries stalking deer in these same hills.

I mentioned to him that many a Marine would do well to follow him on a stalk, learning more in a single afternoon than in many months of training.

'Aye, that's an interesting thought,' he replied, chuckling drily. 'I sometimes tell my clients that they should imagine the stag has a rifle, too. That generally keeps their backsides down on the final approach.'

Dave halted ahead of me, and I stood stock still. He slowly cupped his hands to his face. He seemed to be sniffing the air, picking up the merest hint of the bestial stench of a deer. I stood like a statue, one foot raised, as I awaited the next instruction, lost in admiration.

Dave turned back to me, the rolled cigarette he had just lit glowing between his lips and billowing clouds of white smoke. There was a look of some surprise on his face. 'What are you doing back there, Monty?' he asked in a perfectly normal voice. 'And why are you standing like that? We've got at least a mile before we need to start properly stalking.'

Soon we were on top of the ridge that would lead us to the final stages of the stalk, and Dave's demeanour changed dramatically. As the wind buffeted and bullied us with percussive thumps, he crouched down and beckoned for me to do the same.

'Right, this is where it really begins. Stay behind me at all times, make sure your clothes aren't flapping in the wind, keep your hat on – it breaks up the outline of your face – and follow my instructions. It's a tricky old day, but if we both concentrate, we may just get you a stag. Good luck – let's go and see what's below the ridge.'

If I thought he had been moving quietly before, now he became wraithlike, with each movement calculated and frequent pauses to check the wind and scan ahead. After several hundred metres, when we had dropped off the ridge and were moving along the steep shoulders, he waved me forward.

He whispered conspiratorially. 'A group of stag next to that re-entrant,' he said, gesturing to a point several hundred metres ahead, 'and one of them is in very poor condition. That's the animal we're going to take. There's another group about a hundred metres below them, so this will be a very difficult stalk indeed – the two groups are in different positions in terms of the wind, so if one picks us up it'll scatter the others.'

Looking through the binoculars I could just make out a group of dun-coloured bodies next to a rocky ridge that meandered down the slope. My pulse quickened, a battery of conflicting emotions vying for attention. I could clearly make out the stag that was our target, a ragged-looking creature that, even at the end of what had been a relatively mild summer, looked lean and drawn.

We crept ever forward, frequently on our bellies, pausing for minutes at a time, always in the low gullies and marsh. On the occasions we spied the deer, ever closer as the minutes turned into an hour and beyond, we could see the flick of their ears and the raising of their heads to sniff the wind.

'They know something's up,' whispered Dave, sotto voce. 'They'll have picked up our scent in this wind.'

I had given Dave an impossible task on this day of all days. The wind was being whipped and spun around the ridges, valleys, and cols of the escarpment. It constantly changed direction, and yet I was demanding that he lead me to within a hundred metres of two groups of stag spread on the hill before us, both in the eye of a different set of gusts and breezes carrying all the information they required to detect our presence.

Nonetheless, with pause after pause, and an infinitesimal set of slow movements, he drew me ever closer. Throughout I could feel my senses at their sharpest, returning to the most basic skills all of us possess. Deep, deep within me, way beyond the civilised veneer of the conservationist and biologist, a distant thread of DNA came gloriously alive and rejoiced utterly in the hunt.

Finally we were a mere eighty metres from the stag. Dave rolled to one side and slowly inserted five rounds into the breech of the rifle, each moving in with an oily snicker. He slowly handed me the rifle, both of us just below the flickering grass of a gentle rise ahead. I crawled to a position where the stag was visible against the green rise of a mossy bank, and slowly placed my eye against the sight.

The stag leapt into focus in the crosshairs, head down as it fed, the spread of its antlers brushing the long grass.

Dave placed a hand on my shoulder, his voice so soft it was part of the wind itself. 'Not yet, only when I tell you.'

The stag continued to move down the hill, still feeding, but now there was a change in the group around it, an urgency and tension lacking only seconds before. One of the other animals in the herd had picked up on something, possibly the click of the safety catch. The stag in my sights stopped, and raised its head, a picture of stillness and yet quivering with explosive potential energy.

'OK, take the shot now,' Dave whispered.

As the crosshairs steadied on the chest of the deer, as my

finger tightened on the trigger, as the levers and springs within the weapon began to move and creak, so the stag moved and the hand was on my shoulder again.

'Wait, the shot has to be clean.'

I eascd the pressure on the trigger, and as I did so, the stag took flight, going from a standing start to a gallop in the space of a single stride. Gone – over the ridge in the blink of an eye, an image of speed and flight that had been a second from death yet now appeared the very image of life.

I dropped my head from the sight, letting my breath out in an explosive gasp. I felt completely drained, spent at the release of the tension of the hunt. It had been an hour of brittle tension, and my senses hummed and quivered under the strain.

Dave patted me on the shoulder, shrugging and smiling at the same time. He congratulated me on the stalk, as I did him. It had been a day on the hill when I had learned so much, when I had thrilled at the slow march towards an ancient prey, and when I had held an animal's life in the crosshairs but fate had decreed I should not pull the trigger. I dared not whisper it to Dave, but for me the conclusion to the day could not have been finer.

13 September

The Goodbye Party

It was a source of some amusement around the village that I would soon have to take the pigs to the slaughterhouse. The widely perceived wisdom was that I would blubber uncontrollably, that my fingers would have to be prised from the coarse hair of Gemma's shoulders, and I would be found staring morosely into my pint that evening, mad with grief.

I had undoubtedly become attached to the pigs during their time in the bothy. However, I had also learned a great deal from Keith and Rachael.

'You do your very best when you raise an animal, Monty – that's all you can hope for,' Keith told me one afternoon as we leaned on the rough stones of the enclosure. 'Then you make sure they die quickly and efficiently, and you make the most of the meat. That's the crofting way.'

I surprised myself when the time came to take the pigs to the slaughterhouse in Dingwall. I felt nothing but a sense of purpose as I guided Gemma and Doris into the trailer, and a similar detachment when I dropped them off two hours later. I permitted myself a quick final scratch behind Gemma's ear before leaving, but by then she was already nosing her way through the straw on the holding-pen floor, quite oblivious to my presence. The end of the pigs would be very quick indeed. I knew that Gemma and Doris were bright, clever, adaptable, powerful animals, full of character and humour. I also knew – without wishing to sound callous – that they were going to be absolutely delicious.

Keith came to take away the rest of the Soay sheep, the final sound I heard being the familiar skitter of cloven feet on the metal floor of the trailer as it rocked up the steep track to the main road. They would be returned to his croft on Skye, back to a remote hill on a wild island, where the waves crashed into the cliffs and sea eagles wheeled overhead in the hunt. Back to where their kind truly belonged.

With their departure, the bothy was suddenly stripped of life, the gates to the enclosures swinging on their hinges, the walls silent and the air empty. I awoke for the last few mornings to a deep silence from the fank. Far from feeling liberated by the release from the tyranny of the daily routine, I felt a great sense of loss. The bothy seemed to me to feel this as much as I, the wind moaning round the empty walls and sighing over the open fields where once the sheep grazed.

My time on the west coast was over in all but name. However, there was one final, gigantic hurrah left: the great party to celebrate everything that had been Beachcomber Cottage.

*

As the day for the party approached, it seemed that everyone in the village had heard and was coming along. I simply had to say 'And are you coming to . . .' to be interrupted in full flow by a laconic 'Aye, I'll be there. Would ye mind if I brought a few friends along as well?'

It was slowly dawning on me that this was no ordinary get-together. For me, the very essence of Beachcomber had been the local people, a constant source of advice, comfort, hospitality and friendship. The bothy had not triumphed because of me – in fact, there were many, many times that my ham-fisted efforts had created more problems than they had solved. It had worked because every time I tilled the vegetable patch there seemed to be someone leaning on the wall offering a few tips. Every time a pig did something bizarre and unexpected, there was a friendly word to put things right. And every time I stared miserably into the middle distance in the pub, lost in my thoughts, an arm would drape around my shoulders and a whisky would magically appear before me.

One of the joys of Applecross is the irresistible momentum of the collective and, once rumours of the party had begun to circulate, all manner of villagers began to offer their assistance. The quagmire that was the track to the bothy presented a considerable barrier to getting the marquee on site. ('Ach, ye cannae do this with just a few small tents, Monty – ye'll need the big top!' Alfie had told me dismissively one night. 'Leave it to me.') Judy swung into action to book the entertainment, finally announcing triumphantly that she had secured the services of the Bigfield Blues Band, a

273

collection of bearded, pig-tailed maestros, whose reputation for electrifying any gathering stretched the length of the west coast. A local pig-roast specialist was engaged to cook Doris – creating the rather unusual prospect of me actually sinking my teeth into her, not the other way round.

On the day of the party, a fleet of four-wheel drives thundered over the horizon towards the bothy. Soon there was a group of smoking locals peering at the muddy tyre tracks by the gate, hatching elaborate plots to ensure that the generator could be delivered. This was a serious machine, a clattering yellow box the size of a pool table that weighed several hundred kilos. I hovered on the edge of this conversation, occasionally remarking that I already had electricity in the bothy with the wind turbine; couldn't they just plug into that? I was dismissed in a haze of cigarette smoke – the prospect of deeply grooved tyres spraying mud in all directions and the hefting of gigantic bits of machinery was all too enticing.

Dave Keeper solved all the problems of delivery by fetching the Argo, in which he trundled happily back and forth carrying ludicrous loads, which the small army of helpers swiftly built into an impressive party venue. The snow-white marquee dwarfed the bothy completely, flapping and billowing in the gentle afternoon breeze.

The warmth of the afternoon gave way to the chill of a tranquil autumnal evening, and a steady stream of guests tramped muddily into the marquee. Peter and Jackie appeared, with Peter carrying a bottle of his famous sloe gin. I tucked it away in the bothy, to be drunk in the midst of winter back in Bristol, a heady taste of another time and place. Keith and Rachael turned up with the family, the kids immediately charging off with Reuben to the beach. Mike and Linda wandered over to say hello, her looking impossibly glamorous, him hollow-eyed and mumbling as he had just returned from an expedition in British Colombia, travelling through Heathrow, Inverness and the Bealach.

'I've been awake for forty hours, you know,' he said. 'This better be a humdinger of a party, Monty, or I shall want my money back.'

He needn't have been overly concerned, as his first pint threw him straight into a jetlagged, burbling haze, and he became quite the life and soul in a detached sort of way.

Dave Keeper and Tim were deep in conversation at the makeshift bar, a conversation that swiftly drew in Andy. The topic of the moment was the top five films of all time, a subject about which Tim had strong opinions, being singularly well read and versed in the arts.

'You have to go a long way to beat the popular classics of course – *Jaws*, *Star Wars*, etc. But then again there are timeless movies such as *Breakfast at Tiffany's*, which would always muscle their way in,' he said, looking thoughtful.

'I thought *Convoy* was pretty good,' said Andy innocently. 'That's probably one of my top five.'

Tim went ashen with indignation. 'The glorious pantheon of the cinematic art form cannot, I'm afraid, have *Convoy* at its head. I simply will not permit that to be in your top five.'

Andy was delighted at this development. 'It's my top five,' he insisted. 'I'll have what I like. I thought *Smokey and the Bandit* was pretty good too – number three, I'd say.'

I left them with Tim bouncing up and down vigorously on the spot, incandescent with frustration.

By now the band had arrived and set up their instruments at the head of the marquee. They had taken the precaution of bringing their own crowd with them, a group of inebriated youngsters who had begun to dance before the music had even begun.

'Ahem . . . sorry about that, Monty,' said Martin, the band leader. 'They follow us wherever we play and seem to really enjoy the music, so we quite like having them around.'

I pointed out that they were younger than the rest of us, more attractive, better dancers, and they had brought their own booze. They were more than welcome to stay.

The band struck up, the music echoing around the natural amphitheatre of Sand Bay, and soon the marquee was a heaving mass of sweating dancers, children and dogs. The band sat beaming at the head of the marquee, strumming, drumming and jigging, a catalyst for chaos.

I retired to refuel between songs. Dave Keeper, as was his custom, was overseeing the structural integrity of the bar, ensuring that it did not topple over at any time by securing it with one elbow and his body weight. He nodded laconically as I approached and wordlessly leaned over to pull a pint, placing it before me as I arrived. We chatted briefly, and then he suddenly straightened, glancing into the bay and removing the cigarette from his lips.

'Well, I'll be. There's a basker right off the rocks.'

After six months of scanning the channel for a basking shark, there, in my final few days in the bothy, was a massive dorsal fin and tail moving only metres offshore, creating the gentle V-shape of an oily wake. The shark was directly in front of the marquee, in shallow water only a metre or two deep. The surface of the water was burnished with the first touch of moonlight, making the fin appear ghostly and surreal.

The band trailed off at the shout that went up from the dancers, word spreading instantly that we had been joined by an ocean giant. There was an exodus out onto the grassy shore, where we all stood in a silent line, as if in homage to the colossal animal that sculled its way slowly around the bay before disappearing into the depths of the open channel.

The band struck up once again, breaking the spell, and the dancing resumed. At last the spit roast was ready, with the lid opened to reveal the deep brown of the cooked pig, the crackling shining in the moonlight. Tiny rivulets of grease tumbled down the flanks, and the first knife cut revealed the delicate white flesh beneath, moist and steaming, flaking into a waiting dish. The smell was overwhelming, a sweet odour that swept through the marquee.

Soon a queue formed next to the pig, with Reuben joining in at the back, shuffling forward and trying to blend in. Jackie – a vegetarian for twenty years – buckled under the sight and smell of the pig, tucking into a pork roll with carnivorous relish, giggling in delight as I glanced at her in surprise, grease shining on her chin.

I had set up a small video camera in the extension with a view to the locals recording their thoughts and final messages to me. When I reviewed the footage the next day, I was treated to a series of shots of bottoms filmed from various angles and messages of dubious quality delivered by inebriated friends. The one crime on the west coast is to take yourself too seriously, and I tucked the tape away for the day when my ego may be in need of restraint.

I had decided several days before that I would auction my possessions from the bothy to raise money for a local charity. The auction was a triumph, a measure of the generosity of the village and the vast flagons of ale that had been consumed. Bidding wars erupted for stained tea towels, a pin cushion went to burly fishermen for an absurd sum and soon every stick of dilapidated furniture was gone. Alfie – a tireless fundraiser – sidled up to me with the widest of smiles.

'Well done, Monty boy, a fine effort,' adding ominously, 'I think we can do better than that, though.'

The next time I saw him, he was waving at me from the front of the crowd, auctioning off a gigantic bottle of whisky that I had rather planned to take home. He winked at me and managed to get the price up to well over a hundred pounds, pig farmer Davy Seal waving his skinny arms in a frenzy of largesse only to look vaguely stunned when his final offer was accepted.

When the party was at its height, I moved quietly into the shadows and headed across the beach, quite alone in the silver light of the moon. I climbed the dune, wheezing and panting in the darkness, until the twisted shadow of the rowan appeared before me. I slumped beneath the overhanging branches, sitting in the inky-black shadow of the tree that had guarded the bothy for nearly two centuries, looking down on the scene below. The marquee pulsed and glowed, whilst the bothy sat proudly on the headland, as alive and vibrant as it had ever been. It looked immaculate and neat on the rocky headland, a grand old lady restored to her prime.

Epilogue

It was the final night in the bothy, the debris of the party long gone and the remains of my possessions packed away. Antje and I had decided that our final night should be a celebration of all the bothy had meant to us, and we ate a delicious meal in the extension, the sea stretching away before us in the moonlight of a perfectly still evening. We retired to bed, both unable to sleep as we listened to the crump of the waves for the final time, overwhelmed by a sense of melancholy at our departure. We chatted in soft voices until midnight, with Reuben snoring contentedly in his bed. Soon the rhythm of the surf worked its magic, and we both drifted off to sleep.

We awoke in the wee hours to a colossal din: Reuben was standing at the front door, barking ferociously, every hackle raised. In the six months that I had lived in the bothy, I had never heard Reuben so agitated, so crazed with aggression and the desire to defend his home. He had a number of different barks, and even on the first night of the storm I had heard nothing like this – a baying that crossed a line between fear and anger. Even when deer had rummaged through the woodpile directly outside, his barking had been excited and alert, not this ancient and disturbing call to arms.

He resisted all our efforts to calm him, circling again and again between the door and the bed, sitting with his back to us to rumble the deepest of growls, eyes fixed on the entrance to the bothy.

I finally walked over to the door to place my ear against it, but I could hear only the gentle beat of the waves. I opened it a fraction, my heart hammering, and Reuben slipped through the gap, a dark shadow ready to do battle. I followed him out, to find him standing in bewilderment, glancing about in the moonlight, tail raised, lip curled and hackles up. A cursory inspection around the bothy, with me fearlessly sending Reuben ahead at every juncture whilst I crouched in the darkness, revealed nothing.

This scenario was repeated throughout the night, leaving us frayed and red-eyed in the morning. Reuben finally calmed as the sun came up, curling up and sleeping as though the night had been nothing more than a dark memory.

I mentioned the events of the evening to James, the brother of the owner of the campsite. He went silent for a moment, then looked at me with a strange half-smile.

'You know, Monty, this is something that we didn't really want to mention, but now you're leaving an' all ...' He trailed off, before beginning to speak once again. 'I think you've been far from alone on that beach. I've heard it myself when I've been camping down there – the sound of a boat being drawn up on the stone slip, footsteps in the darkness, and yet there's no one there. It's just one of those things with Sand Bay.' He looked faintly embarrassed for a moment. 'It's probably nothing – just my imagination, I guess.' His eyes told another story, and I could see a gentle rash of goose bumps spring up on his corded forearms as he spoke.

The people of the west coast of Scotland have an affinity with the hills, streams, trees, sea and animals that some say surpasses normal experience. The older folk would tell stories of those with the second sight – *da shealladh* – or old tales of the evil eye. Even today in certain parts of the Highlands there is a reluctance to accept praise for a beautiful child or a fine animal, in case such praise disguises envy or a curse. Clay effigies – *corp creadh* – were used historically to put the hex on someone, dissolving slowly in water with the unfortunate recipient's death coming when the last fragment of the effigy disappeared. Perhaps the most notorious story locally was of the Annat skull, said to be in use as late as 1900 in nearby Torridon. This was the skull of a suicide victim, and water sipped from it was said to cure epilepsy. The skull belonged to the daughter of the Garve Wizard, who was said to have lured victims to their deaths on the wild rapids of the Black Water.

There were a fair few ghosts said to roam the craggy coast and the shadowy glens. A hundred years ago, the postman who went from Applecross to Lochcarron (and what a round that must have been – once a day over the highest mountain pass in Britain) was said to have seen a woman ahead of him several times, and try as he might he could never catch up with her. The woman was said to vanish into a waterfall at the end of each pursuit. There was also the ghost of a black dog above Ardmore, and a veto on passing along the Shieldaig to Torridon road at night. There were even those who claimed the 'sanctuary' of Applecross actually referred to it being a refuge for murderers and robbers, the inaccessibility of the village meaning a level of security for the less desirable elements of the west coast in centuries past.

Before my arrival here, I had never been one for believing in ghosts. To me, there is a logical explanation for most things. However, James's words simply added to a long-harboured sensation that I was not the only resident in Sand Bay. Perhaps there is a chance – just a chance – that certain things are not entirely rational, even today, when it seems all of the natural world has been broken down into its constituent scientific chunks. I had always felt a benign presence in the bothy, and perhaps a final visit in the small, moonlit hours of the west coast night was a last farewell, a gentle inspection from those whose lives and stories remain forever intertwined in the dark stone of the walls.

*

I wandered the quiet shell of the bothy for one last time, touching the rough walls, leaning over the empty enclosures, straightening the lids of the water butts and closing the gates. Standing inside, it struck me that the building did not feel desolate or empty, it simply felt that it had returned to slumber, quietly waiting for the next breath of life, for the front door to rattle and creak open once again.

I walked outside and called Reuben in from the beach,

feeling a pang of regret that he would not know that this was his final run on the edge of the surf at Sand Bay. He galloped towards me, tongue lolling, his pelt spiky and matted, before leaping onto the back seat of the Land Rover.

The surf cracked and thumped on the wide sweep of beach, a rhythm broken only by the sharp calls of the oystercatchers at the water's edge. The heather and bracken on the hill rustled in the brisk autumnal wind, bringing with it a hint of salt and seaweed along with the promise of sharp winter cold to come. Turning up my collar, I walked towards the car without a backward glance, starting the engine and driving up the hill towards the head of the beach.

I drove up the coast road until I was on the high ground above the beach. Stopping the Land Rover, I looked down on Sand Bay and the cottage for the last time. This rough little building on a wild headland had withstood so much, and now it stood alone once again – quiet and calm, but ready to host life. If it had to wait a while, then so be it. After a century alone and neglected, it seemed to me that the one thing Beachcomber Cottage possessed in abundance was patience.

I slipped into gear, gunned the engine and moved forward, glancing back despite my best intentions. The final image was of the bothy stood in the grey light of the afternoon, neat and compact on the rocky headland, before it was hidden by the great bank of the dune and the dark rocks of the cliff wall as I gathered speed and pulled away.

Acknowledgements

There are so many people who made this venture possible that it is an overwhelming task trying to thank them all. I shall make a valiant attempt though.

Thanks to Dick Colthurst and Andrew Jackson from Tigress Productions, who never gave up on me despite overwhelming evidence to the contrary.

Thanks to my big sister Patsy – the talented member of the family – who, despite spending only a few days at the bothy, produced the brilliant sketches within this book.

Thanks also to Martin Pailthorpe, for his relentless creativity, enthusiasm, outstanding cooking and good company.

To Matt, Cat, Ian, Annie – I can think of no finer group of people with whom to share this experience.

To Amy Twomey who had the worst job of all.

Particular thanks to Jo Sarsby who has worked tirelessly on my behalf over the last three years, and Julian Alexander who coached and cajoled a budding author thoroughly intimidated by the task he faced.

To Albert DePetrillo, for allowing me to send him angry emails and always replying in good grace and humour.

To Antje, for always being there for me, even when tagging annoyed penguins thousands of miles away.

Thanks to Keith and Rachael Jackson, for their endless patience, good humour and whisky.

To the Applecross Trust – and Archie MacLellan – for allowing the entire project to go ahead.

To Mike and Sue Scott – great hosts, great diving.

To Mike Dilger, for writing *Nature's Top 40: Britain's Best Wildlife* – my absolute bible for my time in Scotland.

Finally – and significantly – thanks to the people of Applecross, for the warmth of their welcome.

Picture credits

All photographs copyright Monty Halls, except the following.

Martin Pailthorpe: plate section 1, p.1 (top), p.3 (middle), p.5 (bottom), p.6 (bottom), p.8 (middle); plate section 2, p.1 (top), p.8.

Ian Hay: plate section 2, p.7 (middle, bottom left and bottom right).

All illustrations by Patsy Breach.

Index